Acknowledgements

THE AUTHOR WOULD like to thank all those who contributed their time and expertise to the development of this book, in particular Jane Sellars, art historian, curator and author, for her inspired and invaluable picture research. The author is grateful to Eddie and Chris Hewison for their rare combination of music and train expertise, and to musicians Mike and Helen Limer for their help with languages, proof reading and music notation. Thanks are also due to Peter Owens for his translations.

Contents

Introduction   I

RAILWAY MUSIC IN THE UNITED KINGDOM IN THE NINETEENTH CENTURY

1   Broadside ballads, navvies on the line   3
2   Music hall and the railway   27
3   Railway works bands, choirs and musical societies   45
4   Gilbert and Sullivan and the railway   63

THE COMING OF THE RAILWAYS TO EUROPE

5   The coming of the railways to Austria, the Strauss family and railway music   75
6   The coming of the railways to Scandinavia, Hans Christian Lumbye and others   89
7   The coming of the railways to France, Charles-Valentin Alkan and Hector Berlioz   97
8   The coming of the railways to Belgium, Gioachino Rossini   109
9   The coming of the railways to Russia, Mikhail Glinka   117

RAILWAY MUSIC IN EUROPE AND THE USA IN THE TWENTIETH CENTURY

10	Three British railway pieces from the 1930s, *Night Mail*, *The Way to the Sea* and *Coronation Scot*	127
11	Railway music in Paris between the wars	137
12	Four pieces by Percy Grainger and Charles Ives	151
13	Railway music after World War II	161
14	A more experimental approach	173

RAILWAY MUSIC IN NORTH AMERICA IN THE NINETEENTH AND EARLY TWENTIETH CENTURY

15	The coming of the railroads to North America: work songs, hoboes, gospel music and the blues	187
16	Heroes and villains of the American railroads: John Henry, Casey Jones, Railroad Bill and Jesse James	207
17	Trains, lines and wrecks on the early American railroads	221

POPULAR MUSIC

18	Sounds of the railroad in boogie-woogie, bluegrass, blues and jazz	239
19	A medley of popular song	255

INDEX

General Index	281
Classical Music Index	287
Popular Music Index	291
Index of Song Titles	297
Select Bibliography	303

Introduction

WHEN THE STOCKTON & Darlington Railway opened in 1825, it was the first steam-powered railway to carry passengers. Since then there has been no shortage of music connected with trains and railways: orchestral pieces and popular songs describing railway journeys; those that celebrate the opening of a new line; work songs and blues describing the hardship of building the railroads, even the first use of sampled music used railway sounds as its source. From the pastoral serenity of the Flanders and Swann song 'Slow train' to the shrieking horror of holocaust trains in Steve Reich's *Different Trains*, the railway has inspired countless pieces of music. So what is the appeal which has attracted so many musicians? Is it the relationship with time where the train moving through the landscape speaks of the physical and metaphorical power of the railway to connect people and places? Or is it simply the attraction of the evocative sounds, the roaring and wheezing of the steam train, the shriek of the whistle and the clattering track rhythms that are so easily recognisable and adaptable to change? Perhaps composers are in part attracted by the parallels between a train journey and a composition. Both have a beginning, middle and an end and take place across time. It is not difficult to see how the sometimes laboured departure of a steam engine can be performed as a slow introduction, building up to a quicker development, then slowing down to a halt as the passengers reach the end of their journey. It is an easy model for composers to follow and there are many examples of this blueprint to be found from the 'The little train of the Caipira' by Villa-Lobos to Duke Ellington's 'Daybreak Express'.

The earliest train songs can be traced back to broadside ballads; rhyming ballads sold on the street to spread the news. Songs such as 'Navvy on the line' tell of the building of the railways; and railway openings are celebrated in, for example, the 'Glasgow and Ayr Railway'. Some ballads are the bearers of bad news, telling of the many tragic accidents brought about by the railway, some welcome the railway, others criticise it. In the United States, the coming of the railroad offered a promise of escape giving access to employment for farmers and labourers in the South, many of whom became hoboes or migrant labourers and travelled to the North. Their railroad experiences are documented in such early railway songs as 'Hobo Bill's last ride' and 'Rambling on my mind'. The field recordings collected by John A Lomax and his son Alan prove to be a particularly rich source of work songs and the blues. In the 1930s, the two folklorists were sent out by the American Library of Congress to collect recordings. Their tour of gaols in the Southern States where many blues singers (including Leadbelly) were incarcerated provided recordings of 'Take this hammer', 'Midnight Special' and the 'Wabash Cannonball' amongst many others.

The railway had an effect on many aspects of society; it stimulated economic development and had the capacity to change people's lives. In Europe the railway could act as a significant tool in nation building; and for some countries it was an important instrument of military strategy. Across Europe, the opening of new lines was celebrated in music including two of the most well-known railway pieces, the famous *Eisenbahn-Lust-Walzer* (Railway Pleasure Waltzes) by Johann Strauss I which was composed in advance of the much anticipated 1837 opening of the line between Floridsdorf and Deutsch Wagram in Austria, and Hans Christian Lumbye's *Copenhagen Steam Railway Galop*, written to celebrate the opening of the line from Copenhagen to Roskilde. For some time the train was seen as the epitome of modernity. In the early twentieth century the Futurist movement sought to express this in mechanical imagery. One of the most well-known orchestral train pieces, Arthur Honegger's *Pacific 231*, was created in this spirit, evoking raw power and inexorable machine-driven motion. As the twentieth century moved on, this perception of modernity eventually faded and instead steam trains were sometimes used a symbol of nostalgia in, for example, the Kinks' song 'Last of the steam-powered trains' where the train acts as a metaphor for the past. There are many other examples of songs where the train is used as a metaphor, an obvious example being 'Morningtown ride' by The Seekers which uses a train journey as a metaphor for sleeping. In spirituals and gospel music a railroad trip is often used to represent the journey through life with heaven as its final destination, some songs such as 'Life's railway to Heaven' include a warning that any departure from righteousness would lead elsewhere.

There is much to be learnt from the lyrics of train songs. Train journeys are a common theme, often linked to notions of connection and isolation: songs about the first train and the last train; leaving and coming home; and separating lovers and bringing them back together again. In some ways lyrics can act as social documents which capture historical experience, whether it be the humorous songs about the railway found in British music hall which look at life from a working-class perspective, or the voice of resistance heard in work songs and the blues which share aspects of the black experience. Some songs embody political values and others are overtly political, not least Hugh Masekela's powerful song 'Coal train' with its grim documentation of the trains that transported migrant workers to and from the gold mines in South Africa.

The book reveals some interesting insights into the relationship between the musicians and the railways. The Italian composer Rossini hated trains. This was probably a reaction to a catastrophic train journey from Antwerp to Brussels in 1836 where he was so unnerved by the experience that he fainted, was ill for several days afterwards and swore that he would never set foot on a train again. Nevertheless this did not prevent him from writing *Un petit train de plaisir comico-imitatif* (A little train of pleasure). Similarly, the composer Percy Grainger wrote *Train Music* as a response to a journey in 'a very jerky train going from Genova to San Remo'.

Those musicians who loved trains, however, are to be found more frequently amongst the composers and songwriters featured in this book. When Benjamin Britten was working on *Night Mail*, he enjoyed spending evenings at Harrow station listening to the trains arriving and departing. Ex-punk Captain Sensible of the Damned once revealed that 'The Damned use touring to pursue our train obsessions' and visit 'as many steam preserved lines as possible'.[1] Tom Waits is similarly obsessed and collects the sounds of trains on tape. Several composers used the train as a sanctuary and a place to compose when on tour; Duke Ellington rented his own Pullman train car so that he would have a place to eat and sleep when touring in segregated towns. The American Charles Ives enjoyed his two-hour daily commute and it was one of these journeys which inspired his piece 'From Hanover Square North, at the end of a tragic day, the voices of the people again arose' recalling his experience of communal singing at the station the day that the news broke of the sinking of the RMS Lusitania.

In all the book describes over 50 pieces of classical music and covers over 250 songs, recordings of all of these can be found on the internet. Given the huge number of pieces of train music, particularly songs, it has, of course, been impossible to write about each of them. Some of these have been captured by title only with several pages of lists of classical pieces, jazz numbers and popular songs. There are also, of course, many instances of railways and music that can be found in film and television, enough to fill another book in fact, but these have not been included.

Endnotes

1   https://thequietus.com/articles/03195-the-damned-s-captain-sensible-on-why-he-likes-trains

*Railway music in the United Kingdom
in the nineteenth century*

1 Broadside ballads, navvies on the line

THE OPENING OF the Liverpool and Manchester Railway in 1830 signalled a huge landmark in railway history; it was the world's first inter-city railway, it carried both passengers and goods, and it travelled in both directions. Designed and built by George Stephenson (1781–1848), the Liverpool and Manchester Railway was the culmination of decades of experimentation with tracks, engines and locomotives. Its construction is widely recognised as the beginning of the development of the railway network, not just in the United Kingdom but across the world.

Trains were faster and cheaper, passengers and goods could now travel between Liverpool and Manchester quicker than ever before. They aided economic and social development, and by changing concepts of time and distance, they altered people's horizons and increased their opportunities. In effect the new railway sparked a revolution in trade and travel. The immediate financial success of the Liverpool and Manchester Railway, along with new developments in locomotive design, led to a rapid expansion in railway construction and the UK network developed very quickly. This boom in railway building is often referred to as 'railway mania'. In 1844 there were 104 railway companies, and then over the next six years another 110 were added.[1] Between 1833 and 1843, 2,300 miles of track had been built.[2] By 1870, with over 15,000 miles of track, the main line network was almost complete, and by 1912, there were 23,441 miles of track open to traffic.[3]

The massive undertaking of building the railways fell to huge numbers of itinerant workers known as navvies. Their travails are recorded in some of the earliest railways songs in the form of broadside ballads, long rhyming ballads written in part to spread the news, and in part to entertain.

The connection between music and the railways can be traced back to their construction with some of the earliest music being found in broadside ballads: songs such as 'The navvy boys' tell of the building of the railways; the highs and lows of 'railway mania' are recorded in such broadside ballads as 'The rail, the rail'; and railway openings are celebrated in, for example, the 'Newcastle and Carlisle Railway'. Some songs welcome the railway. The 'Western Railroad' of 1863 celebrates the new railway coupled with the downfall of the stage coach business. Others criticise it, such as 'Shillibeers Original Omnibus versus the Greenwich Railroad' which instead extols the virtues of travel by omnibus. There are also many ballads that are bearers of bad news, news of some of the many tragic accidents brought about by the railway.

Broadsheets or broadsides (single sheets printed on one side) were among the earliest products of the printing press and date from the sixteenth century. They were sold on the street and were designed to entertain and inform the masses, spreading the news, whether of national events or local scandal. Such street literature included ballads (sometimes known as slip songs), and chap books which were made by printing a large sheet of paper and folding it into a small booklet. Broadside ballads covered a wide range of subjects from murders and executions, to deaths of the famous and local gossip. The themes speak of contemporary issues, struggles and poverty, political events and communal tragedy as well as local tittle tattle and bawdy stories; as such they are a valuable and entertaining social record. Rapid technological advances in steam locomotive design in the late 1830s led to a huge expansion in railway building which accelerated in the 1840s and 1850s. The world was rapidly opening up; eventually anyone who could afford the train fare became free to travel longer distances for work or leisure. Naturally railways were a hot topic.

The ballads were printed on cheap paper sometimes with a woodcut or engraving heading the sheet. Music notation was rarely included; the ballad writers assumed that their customers had a repertoire of popular melodies which had been passed on orally. So instead of notating the melody, a well-known tune was often suggested on the broadsheet, hence many different ballads were sung to the same, usually repetitive and rhythmic, melody. At other times melodies were written especially for a new ballad – in an earlier century the composer Henry Purcell was a prolific writer of ballad melodies. The ballads were sold in large numbers on the street by travelling ballad singers who would perform them on street corners, town squares and fairgrounds drawing a crowd of potential customers and singing them through a few times giving the buyers the opportunity to learn the tune. The peddler would sing them, stopping before the last verse so that 'they didn't give away the end of the story'.[4] The ballad was a 'successful product tailored to accommodate the largest possible readership. At the price of one penny, ballads were affordable to many of the poor, and when sung aloud they could be experienced even by the illiterate.'[5]

Broadsides would be pasted on top of each other onto the walls of homes and ale houses, church doors and other public places. They were 'sung, read, memorized, collected, quoted, copied—or met more ignominious ends as kindling for a pipe or paper for the privy house.'[6] Because of their ephemeral nature and the fact that the paper was thin and cheap, many did not survive. Nevertheless, some were collected and can still be found in libraries in the UK and North America.[7]

Here is a selection of broadside ballads about the railways. There is a degree of overlap between broadside ballads, folk songs and songs found in the music hall. Many of the early ballads were derived from folk songs previously transmitted orally, and some broadside ballads became so popular that they were taken up by music hall performers.

50 broadside ballads about the railways

SONG TITLE	DATE
Building the railways	
Bold English navvy	1839
The navvy Boy	1839-41
Paddy on the railway. *Come landlord fill the flowing bowl*	1840
Paddy works on the railway. *In eighteen hundred and forty-one*	1840s
Hull and Holderness Railway, 20th. October, 1853	1853
Paddy on the railway	1858-1885
Paddy on the railway. *Paddy one day from Greenock Town*	1860
Battle of the navvies	1864
Irish harvestmen's triumph	1867
Navvy on the line	Unknown
The Irish navigator	Unknown

The navigators	Unknown
Dashing navigator, A new comical song called the	Unknown
The navy boys	Unknown
Railway mania	
The rail, the rail	1845
Railway mania	1846
The Railway King	1849
Railway openings	
Newcastle and Carlisle Railway	1835
Jim Crow's description of the New Greenwich Railroad	1836
A new song on the opening of the Birmingham and Liverpool Railway	1837
The opening of the new railway	1837-38
The railway	1838 -
The New London Railway	1839
Glasgow and Ayr Railway	1840
Pennyworth of fun; or, Opening the Oxford Railway	1852
Impact of the railway	
Liverpool improving daily	1828
Manchester's an altered town	1830s
Liverpool's an altered town	1830s
Shillibeers Original Omnibus versus the Greenwich Railroad	1838
The Great Western Railroad or, the pleasures of travelling by steam	1840
Wonderful effects of the Leicester Railroad	1840
The railway	1850
Kendal Fair	1850
Western Railroad	1863
Glasgow is improving daily	1860s
Halifax, Thornton and Keighley Railway	1870s
Accidents	
The falling of nine arches and fifteen lives lost at Ashton, April 19th 1845	1845
Awful catastrophe at the Clayton Hill Tunnel on the Brighton Railway, on Sunday August 25th 1861	1861
Awful railway accident between Peterborough and Huntingdon	1876
Lines on the railway collision at Burscough Junction	1880
Awful railway accident. breaking of a bridge over the River Tay	1880
Railway outings	
Johnny Green's trip fro' Owdhum to see the Manchester Railway	1842
Kendal Fair	1850

The Cockneys trip to Brummagem	Unknown
Riding in a railway train	Unknown
Railway workers	
Jessie at the railway bar	1884
Signalman on the line	Unknown
Shunting pole inspector	1898
The Muddle Puddle porter	Unknown

These broadside ballads come from four collections: the Bodleian Library, *allegro* Catalogue of Ballads, the University of Oxford; the Sir Frederick Madden Collection, Cambridge University; the National Library of Scotland, English Ballads Collection; and the website songsfromtheageofsteam.uk

Navvies building the railways

The word 'navvy' was used to refer to those men building the railways in the nineteenth century. It comes from 'navigator', the name given to the canal builders of the eighteenth century.[8] There was plenty of employment for navvies from the beginnings of the railway until the onset of World War 1, not just constructing the first lines but also making improvements to existing lines. It is estimated that there were 200,000 men working on the construction of the railways during 1846-7, the period of 'railway mania' - and they were all men. Women often joined their husbands on the shanty encampments built next to the construction sites. They did not find labouring work there although it should be noted that there were three women listed as 'railway labourers' in the 1851 census and in the 1850s Elizabeth Ann Holman worked on the construction of the Great Western Railway in Cornwall as a navvy by 'masquerading as a man'.[9]

It would be wrong to give the impression that the navvies were merely labourers, although in later years this is often how the term has been used; many of the men were skilled as miners, masons, carpenters and blacksmiths, for example. As Simon Bradley writes in *The Railways: Nation, Network and People*, 'Drunk or sober, the men faced deadly risks at work'.[10] The work was arduous and dangerous: navvies excavated, blasted and tunnelled their way across the countryside; 'shifting 20 tons of earth was a normal day's work'. Many navvies were killed by explosions and collapsing tunnels and 'three accidental deaths per mile was considered an acceptable average'[11]. However, the work was well paid, particularly in comparison with the wages of an agricultural labourer, the employment that many of the navvies would have done before the days of railway construction.

A gang of navvies near Haddenham, Buckinghamshire take time off from their work on the Great Western & Great Central

Photo by S W A Newton, 1903. Reproduced with permission of the Record Office for Leicestershire, Leicester & Rutland

Navvy on the line

I am a Navy[12] bold that has tramped the Country round sir
For to get a job of work where any can be found sir
I left my native home, my friends and my relations
To ramble up and down the Town, & work in various stations.

Chorus
I am a navvy don't you see, I love my beer all in my prime
Because I am a Navvy that is working on the line

I left my native home on the first of September,
That memorable day I still do remember
I bundled up my [illegible], put on my smock and Sunday cap sir,
And wherever I do ramble, folks call me Happy Jack sir.

I have got a job of work all in the town of Bury
And working on the line is a thing that makes me merry
I can use my pick and spade, and my wheelbarrow;
I can court the lasses too, but never intend to marry.

I worked there a fortnight and then it came to pay day
And when I geet my wages I thought I'd have a play day
And then a little spree in Clerke Street went quite handy
And I sat me down in Jenkinson's beside a Fanny Brandy

I called for a pint of beer and bid the old wench drink sir
But while she was a drinkin she too at me did wink sir
Well then we had some talk, in the backside we had a rally
Then jumped over brush and steel & agreed to live tally

They called for liquors merrily; the jugs went quickly round
That being my wedding day, I spent full many a crown, sir
And when me brass was done, old Fanny went a cadging
And to finish up me spree, I went and sloped me lodgings

Oh now my chaps, I'm going to leave the town of Bury
I'm sorry for to leave you chaps, for I've always found you merry
So call for liquors freely and drink away me dandy
And cry out here's health to Happy Jack, and Fanny Brandy

Terry Coleman gives a fascinating and detailed 'History of the men who made the railways' in his 1965 book *The Railway Navvies*. In many ways the broadside ballad 'Navvy on the line' encapsulates the life of a navvy as described by Coleman. It tells the tale of the itinerant existence of many navvies where they tramped around the country in search of railway employment. This hard-drinking lifestyle is introduced in the first verse and the chorus. The song is not dated but several references are made to the town of Bury. The first line to Bury was built in 1846 on the East Lancashire Railway running north to south. It was an extraordinary feat of engineering. There was a shortage of labour in the summer of 1845, men were asked to work on Sundays and consequently seven labourers were charged with breaking the Sabbath. In 1848 the line between Bolton, Bury & Heywood running east to west opened on the Lancashire & Yorkshire railway, another challenge with the last three miles taking two years to complete.[13]

In the second verse we are given some details of the clothes that the navvy wore, 'the smock and Sunday cap'. This is corroborated by Coleman when he writes

> The dress too was distinctive. They wore moleskin trousers, double-canvas shirts, velveteen square-tailed coats, hobnail boots, gaudy handkerchiefs, and white felt hats with the brims turned up. They would pay fifteen shillings, a great price, for a sealskin cap, and their distinct badge was the rainbow waistcoat.[14]

Bradley writes that 'Men even abandoned their old identities, picking up vivid new monikers as they went from place to place'.[15] This is reflected in the line 'And wherever I do ramble, folks call me Happy Jack sir.' We hear of Happy Jack's pay day drinking in Verses 4 and 5 where he has a 'spree in Clerke Street'. As Coleman writes 'The navvies were paid once a month, sometimes not so frequently, and usually in a public house, and then for days afterwards they drank their pay, sold their shovels for beer, and went on a randy.'[16] He explains that a 'randy' was a long drinking bout, often lasting four or five days until the money ran out, 'devoted to a celebration of drunkenness, fighting, poaching, robbery, and, occasionally, high-spirited murder.'[17]

Verses five and six tell of Happy Jack's marriage to Fanny Brandy where they 'jumped over brush and steel & agreed to live tally'. It would appear that such an unofficial ceremony was not uncommon for navvies at the time.

> At Woodhead in 1845, where 1100 men were camped in shanty huts, they even had their own marriage ceremony: the couple jumped over a broomstick, in the presence of a roomful of men assembled to drink upon the occasion, and were put to bed at once, in the same room.[18]

The ballad concludes with a call to cheer 'the health to Happy Jack, and Fanny Brandy.'

Paddy works on the railway

In 'Paddy works on the railway' an Irish navvy describes his itinerant career building the railway in the 1840s.

In eighteen hundred and forty-one
Me cord'roy breeches I put on,
Me cord'roy breeches I put on,
To work upon the railway.

Chorus I was wearing corduroy breeches,
Digging ditches
Dodging hitches, pulling switches,
I was working on the railway.

In eighteen hundred and forty-two
From Hartlepool I moved to Crewe,
And found myself a job to do
A-working on the railway

In eighteen hundred and forty-three
I broke me shovel across me knee
And went to work for the company
On the Leeds and Selby Railway

In eighteen hundred and four-four
I landed on the Liverpool shore
My belly was empty, me hands were sore
With working on the railway.

In eighteen hundred and forty-five
When Daniel O'Connell he was alive
When Daniel O'Connell[19] he was alive
And working on the railway.

In eighteen hundred and forty-six
I changed me trade from carrying bricks,
I changed me trade from carrying bricks,
To work upon the railway.

In eighteen hundred and forty-seven
Poor Paddy was thinkin' of going to heaven
Poor Paddy was thinkin' of going to heaven
To work upon the railway.

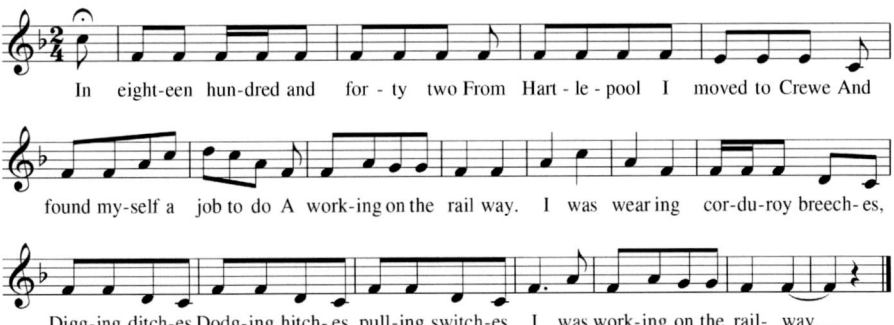

To encourage the widest audience, ballad tunes were usually fairly easy to sing without any awkward leaps and having simple harmony. Notice the way in which the first two lines of the verse to 'Paddy works on the railway' fall mainly on one note and the way in which the lyrics 'digging ditches', 'dodging hitches' and 'pulling switches' repeat the same catchy melodic phrase. The song also uses the same three chords throughout.

The song is sometimes sung to another melody and there are several variants on the lyrics. These include the following verses which were sung in the USA where there were large numbers of Irish immigrants building the railroads. It is included in Carl Sandburg's 1927 collection, *The American Songbag*.[20]

Poor Paddy works on the railway

Oh in eighteen hundred and forty-three
I sailed across the sea (twice).

Chorus
To work upon the railway, the railway,
I'm weary of the railway;
Oh poor Paddy works on the railway.

Oh in eighteen hundred and forty-four
I landed on Columbia's shore (twice).

Oh in eighteen hundred and forty-five
When Daniel O'Connell he was alive (twice).

Oh in eighteen hundred and forty-six
I changed my trade to carrying bricks (twice).

Oh in eighteen hundred and forty-seven
Poor Paddy was thinking of going to Heaven (twice).

Objections to proposed railways

In 1844, the Kendal and Windermere railway line was proposed. The then poet laureate William Wordsworth was outraged, seeing this as a violation of the Lake District, the place where he lived. He wished to preserve the rural beauty of the area and launched a literary campaign against the line. He wrote in protest to William Gladstone, at that time President of the Board of Trade, on 15 October 1844 asking that 'When the subject comes before you officially, as I suppose it will, pray give it more attention than its apparent appearance might call for....'.[21] He backed up his plea by enclosing an anti-railway sonnet: "Is there no nook of English ground secure from rash assault?

> And is no nook of English ground secure
> From rash assault? Schemes of retirement sown
> In youth, and 'mid the busy world kept pure
> As when their earliest flowers of hope were blown,
> Must perish; - how can they this blight endure?
> And must he too his old delights disown
> Who scorns a false utilitarian lure
> 'Mid his paternal fields at random thrown?
> Baffle the threat, bright scene, from Orrest head
> Given to the pausing traveller's rapturous glance;
> Plead for thy peace thou beautiful romance
> Of nature; and, if human hearts be dead,
> Speak, passing winds; ye torrents, with your strong
> And constant voice, protest against the wrong!"

Wordsworth also published the poem as a pamphlet and wrote letters to the *Carlisle Journal* and the *Morning Post* arguing that there were no manufacturers, quarries nor a substantial agriculture base to justify the intrusion and, more controversially, that it would open up the area to the poorer classes who would not have the capacity to appreciate its beauty and secluded character. Furthermore bringing many visitors into the district would destroy the beauty they had come to enjoy. His opposition was 'largely ridiculed', although the part of it that he especially objected to was never built, with the line stopping short of Lake Windermere. Nevertheless the Kendal and Windermere Railway Act authorised construction and it opened on 20 April 1847.[22]

Kendal Fair

You servant lads and lasses gay come listen for awhile,
I'll sing to you a verse or two that will cause you for to smile,
Now Kendal Fair is come again both young and old so thrifty,
Will run with glee to have a spree in 1850.

Chorus
So to Kendal Fair let's haste away to see such lots of fun,
The lasses romping with the lads and playing at tiddle de bumb.

From Burnside and Stavely, the people will come
From Ambleside and Windermere by railroad they will run,
From Stonecroft and Under Barrow, they will come with glee,
Poll and Fan swear they'll have a man for to have a spree

There's Greyrig and Tebay, and Orton likewise,
Milnthorpe and Natland, they'll come of every size,
From Kirby Lonsdale they will come in all their frills and ruffles,
At Kendal Fair I do declare they'll have such stunning bustles.

Now when they come unto the fair they'll wander up and down,
They'll take view of every place that is in Kendal town,
Some to copper lone will steer for to get a glass of beer,
But soon from there they'll retire for fear their bobbins should get fire.

Then to the hirings they will go all for to look for places
Some will hire to milk the cows, some to make cheeses,
Some will hire to hedge and ditch, and some to milk and mow,
And Sally Brass she's the lass to milk her master's doodle doo.

There's black eyed Fan with her frying pan will cook your eggs and bacon,
With beef and mutton roast and boil'd if I am not mistaken,
She'll make the puddings fat and good, all ready for the table,
But if you grumble when she's done, she'll black your eyes with the ladle.

So to conclude and make an end I have not detained you long,
I hope there is no one offended that's heard my little song,
So lasses when you are going home pray with the men don't rustle,
For if they shove you in the hedge you are sure to spoil your bustle.

Kendal is a Cumbrian market town in the south of the Lake District. Its station opened on 22 September 1846. The song 'Kendal Fair' dates from 1850 which is evident from the lyrics. It is printed on the ballad sheet alongside a song called 'Exhibition of all nations!' which looks forward to the Great Exhibition of 1851. The lyrics describe Kendal hiring fair and the people attending. The annual fair matched up labourers with employers and hired them for a fixed term. Such fairs were held in many market towns and they secured work and board for both male and female agricultural servants for a year. As well as acting as an employment agency, the events had all the usual trappings and ribaldry normally associated with fairs. The villages listed all lie within a few miles of Kendal within the historical boundaries of Westmorland. Greyrig (Grayrigg) and Tebay have noteworthy railway connections. The population of Tebay, what was once a small village, expanded to over 1000 with the coming of the Lancaster to Carlisle railway. In 1846 it became an important railway junction housing engine sheds and marshalling yards as well as becoming the home base of banking engines that supplied the extra power needed to climb to Shap summit which has a 1:75 gradient.

Regrettably Grayrigg's history is marked by two major train crashes. In 1947, a 13-carriage London Midland & Scottish Railway service from Glasgow Central to London failed to stop at the signals for Lambrigg Crossing and collided with a locomotive. 33 people were injured, three of them seriously. 60 years later in 2007, Lambrigg Crossovers was the site of the Grayrigg derailment, a fatal derailment, the result of faulty points, involving a Virgin Trains West Coast service from London Euston to Glasgow Central.

Shillibeers Original Omnibus versus the Greenwich Railroad

Wordsworth was by no means the only person to object to proposed railway plans. In 1829 George Shillibeer introduced the first horse-drawn omnibuses to London having picked up the idea on a visit to Paris. His omnibuses travelled between Marylebone and Bank, they were drawn by three horses and carried 18 passengers. Until about 1850 the vast majority of people walked to work. Although the omnibuses were slow and quite expensive for most, they were cheaper than cabs and stage coaches, and 'provided the first instance of public transport that made it possible for workmen to live more than walking distance from their work.'[23] So it is not surprising that Shillibeer objected to the plans for the London and Greenwich Railway Company which received parliamentary approval on May 17 1833 (thus dating this ballad to 1834). The ballad 'Shillibeers Original Omnibus versus the Greenwich Railroad' served as a piece of commercial advertising which extolled the virtues of the Shillibeer omnibus.

Shillibeers Original Omnibus versus the Greenwich Railroad

By Joint Stock company taken in hand
A rail-road from London to Greenwich is plann'd;
But they're sure to be beat, 'tis most certainly clear
Their rival has got the start – George Shillibeer

I will not for certainty vouch for the fact
But believe that he means to run over the Act
Which Parliament passed at the end of last year
Now mad null and void by the new Shillibeer.

His elegant omnis, which now throng the road
Up and down every hour most constantly load
Across all the three bridges now gaily appear
The Original Omnibus – George Shillibeer

These pleasure and comfort with safety combine
They will neither blow up nor explode like a mine
Those who ride on the railroad might half die with fear
You can come to no harm in the safe Shillibeer

How exceedingly elegant fitted inside
With mahogany polished – soft cushions - beside
Bright brass ventilators at each end appear
The latest improvement in the new Shillibeer

At the same time it warns of the dangers of travelling by train: draughts, bursting boilers, explosions that might make passengers 'half die of fear'.

These pleasure and comfort with safety combine
They will neither blow up nor explode like a mine
Those who ride on the railroad might half die with fear
You can come to no harm in the safe Shillibeer

Her no draughts of air cause a crick in the neck
Or huge bursting boiler blows all to a wreck
But as safe as at home, you from all danger steer
While you travel abroad in the gay Shillibeer.

The ballad closes with more fulsome praise for Shillibeer's elegant omnibuses and a plea for the new London and Greenwich Railway to disappear.

That the beauties of Greenwich and Deptford might ride
In his elegant omni is the height of his pride –
So the plan for a railroad must soon disappear
While the public approve of the new Shillibeer.

Shillibeer's ballad was written in vain. The first section of the line opened in 1836 running from Tooley Street (London Bridge) to Deptford: it was the first steam railway in the capital; the first to be built specially for passengers; and one of the first lines to offer season tickets, helping to establish the habit of commuting.

In praise of the new railway

There is no shortage of ballads that welcomed the coming of the railway. 'Glasgow is improving daily' describes the demolition of buildings in central Glasgow to enable the construction of the Union Railway. The first section of the City of Glasgow Union Railway opened in 1870 and included the first railway bridge built across the Clyde.

Glasgow is improving daily

Our city is improving
Every day we all do know
So long may Glasgow flourish
And her sons where'er they go

'Liverpool improving daily' similarly welcomes the benefits of the city's redevelopment because of the railway. It includes the lines 'But now tho' roads are all the go / Railways beat them I've a notion'.

Broadside ballads were sometimes re-workings of earlier ballads. This is certainly true of the ballads 'Liverpool's an altered town' and 'Manchester's an altered town'. Both describe the city's expansion and the consequent changes it had brought about. The Liverpool song includes the lines 'There's Gloucester Street and Nelson Street, have had an alteration / They've pulled the most part of them down to make a railway station' similarly the Manchester song proclaims 'There's Newton Lane I now shall name, has had an alteration / They've knocked a great part of it down to make a railway station'. However it is difficult to establish which came first. Both songs were printed by Harkness of Preston. They both mention new police forces; Liverpool's force was established in 1836 and Manchester's in 1839, and both mention Zoological Gardens. Liverpool's zoo opened in 1833 and Manchester's in 1836.

Railway openings

By 1850 over 6000 miles of railway line had opened, increasing year on year until 1900 when there were almost 20,000 miles of line. In the earlier years the opening of a line was a major event and companies celebrated the event with a grand ceremony. Church bells rang, bunting hung in the streets, canons were fired and there was usually a procession of local dignitaries and railway company representatives, an inaugural train journey, and a massive banquet. Celebrations were attended by large crowds providing an ideal market for ballad peddlers who took the opportunity to sell songs especially written for the occasion.

The Newcastle and Carlisle Railway - A new song

On the ninth day of March in the year thirty-five
The railway was crowded with people alive
From Blaydon to Hexham the engines did move
With all the subscribers united in love
In one hour and ten minutes on that noted day
They returned back on the Newcastle railway

The grand locomotives from Newcastle came
How quick is their speed, how great is their fame
The brilliant Comet she could not well lead
For Rapid came in with abundance o' speed
The air it did ring with the cry of hurra
When they came to open the Carlisle railway.

The hills were all clad on the south side of Tyne
To view the procession along the new line
The drum they did beat and the colours did fly
To cheer the spectators as they passed by
The men will rejoice and the women will pray
For all that subscribe to Newcastle railway

The masons they are the first workmen in town
And some by hard labour can earn a full crown
The blacksmiths and joiners all work to their plan
I can scarcely tell you who is the best man
Let none of these workmen have reason to say
They cannot live by the Carlisle railway.

There is Squire Beaumont, for the sake of his heirs
It is well known that he owns fifty shares
Long may he live with his own darling son
So let us praise him for what he has done
He will hear the birds sing in the sweet month of May
When he travels along on Newcastle railway

There is Mr Blackmoor a worthy young man
To forward this line he will do all he can
In two or three years he will finish it well
And make a through passage into the canal
Long may he live and still carry the sway
And set out more work on Carlisle railway

When you see the steam coaches and all things complete
For four or five shillings you may take a seat
You may dine at Newcastle and then take your flight
And sup at Carlisle on the very same night
The new Expedition she will not delay
As long as she runs on Newcastle railway

The cannons were planted upon the low ground
They made all the vallies to ring with their sound
The drums they did beat and the music did play
Before they went back to Newcastle that day
Both the young and the old may remember that day
When they drank success to the Carlisle railway

When you see the waggons move on at full speed
Well laden with liquor, provisions and lead
You may fill a glass with good rum or strong beer
And then drink a health to the head engineer
I hope he will live to see that happy day
When they have completed Newcastle railway

The tune suggested by the broadsheet is 'Patrick O'Neil'. There are several different variants on this tune, two of them are given below. Notice how both melodies fall into short two-bar phrases. The first variant is 12 bars long and the second is 16 bars long, both use repeated phrases. This is typical of ballad tunes and it meant that they could be adapted to different ballad lyrics. In this way the same tunes were constantly recycled and put into different contexts. Ballads were often connected with dancing and both versions of this song below use the rhythm of a jig.

The 60 mile long Newcastle & Carlisle Railway was the first line to be built between the east and west coasts of Britain. The Newcastle-on-Tyne & Carlisle Railroad Company was formed in 1825 and construction began in 1830. The line opened from Carlisle to Blaydon using horse traction in 1834. The company then decided to adopt steam traction and the first locomotive, No. 1 "Comet," was delivered by the Newcastle locomotive builders R. & W. Hawthorn in 1835.[24] The line had 'some notable engineering features: very fine stone bridges at the western end, and the big elegantly handled Cowran cutting, 2,270 ft long and up to 110 ft deep.' Isambard Kingdom Brunel applied for the post of engineer of the line as did George Stephenson but the job, was given to Francis Giles (perhaps not coincidentally a substantial shareholder in the company).[25] Although Giles completed the design he 'neglected his duties in favour of other railways and was eased out in 1834 in favour of his able assistant John Blackmore'. This is no doubt the Mr Blackmoor mentioned in the sixth verse.[26] 'There is Mr Blackmoor a worthy young man / To forward this line he will do all he can / In two or three years he will finish it well / And make a through passage into the canal.' Since 1794 there had been various plans to construct a canal from Tyne to Solway but these had come to little; in 1829 a company was authorised to build a railway as an alternative. It finally reached the canal at Carlisle in 1837 thus creating a connection between the River Tyne and the canal basin at Carlisle. The line opened for passenger traffic on 10 March 1835, but before this it had been used for freight transport and had already shipped over 500 tons of lead by the end of 1834. The Squire Beaumont mentioned in verse 5 was Thomas Wentworth Beaumont Esq. MP who was the owner of the T.W.Beaumont Lead Company.

The 1835 opening of the first section of the Newcastle & Carlisle line was clearly a happy smooth-running celebration, but this was not always the case with such ceremonies. As Jack Simmons writes 'every sort of misfortune might attend these functions' and goes on to cite the 'exceptionally violent storms' that ruined the later ceremonies organised by the Newcastle & Carlisle, and the Newcastle and North Shields Railway in 1838-40 and the collision between trains at the opening of the Durham Junction Railway in 1838.[27]

Holmfirth is a small valleyed mill town in West Yorkshire surrounded by moorland. Its railway station opened on July 1, 1850 serving a branch line of the Huddersfield to Penistone line.

> ...the first Holmfirth branch train was scheduled to leave the terminus destined first for Huddersfield. From early in the morning the church bells throughout the Holme Valley tolled to signal the importance of the day... and the Holmfirth Band played a series of melodies on the station platform. Despite the shortness of the journey a locomotive and 14 carriages left the branch, with the band now placed in two open carriages. They played all the way to Huddersfield ignoring the torrential rain and only pausing when the train ran through the various tunnels encountered en-route[28].

In later years the opening ceremonies tended to be quieter events and fewer in number.

Accidents on the railway

The Office of Rail and Road is unequivocal when it pronounces that during the first decades of the nineteenth century 'our pioneering railway system was unregulated, uncoordinated and to be honest, downright dangerous.'[29] By 1860 certain precautionary measures had been put in place on the railways, notably the creation of the HM Railway Inspectorate by the Board of Trade in 1840. However, although the Board of Trade investigated accidents they did not have much power, they wrote reports and made recommendations but they had no control over the outcome. Rolt in his book *Red for Danger*, an account of railway accidents through the nineteenth century and into the twentieth century, refers to the 1870s as "a black decade in railway history"[30]; it had an average of nearly 41 passenger deaths in railway accidents per year - the highest level reached on the UK network between 1860 and 1914. It was not until the Regulation of Railways Act in 1889 that the Board of Trade was given the power to require railway companies to adopt the block system of signaling along with the use of continuous automatic brakes and the interlocking of all points and signals, the absence of which had resulted in many tragic accidents and fatalities.

Awful railway accident between Peterborough and Huntingdon

You feeling Christians I pray draw near
And a sad disaster you shall hear
That has caused much pity, distress and woe,
Thirteen poor creatures they are laid low
Upon the railway that fatal night
It must have been a heart-rending site
The Flying Scotsman to London bound
Destruction caused and death around

By the Scotch Express, what an awful sight
Thirteen poor souls were killed that night
Their last death cries were heard around
Has(sic) they died on the snow covered ground

She reached Newcastle that fatal train
And was joined by many a well-known name
They little dreaming what was in store
That many friends they would see no more
Mothers and daughters are parted now
The stamp of death is on their brow
In Heaven we hope they'll meet again
Free from sorrow grief and pain

She reached Newcastle that fatal train
And was joined by many a well-known name
They little dreaming what was in store
That many friends they would see no more
Mothers and daughters are parted now
The stamp of death is on their brow
In Heaven we hope they'll meet again
Free from sorrow grief and pain

To Abbots Ripon it dashed with speed
And there a coal train was ahead
Not able to stay the steam in time
It caused destruction upon the line
The carriage into pieces flew
When a second danger appeared in view
'Ere many scarce could draw their breath
Their eyes forever were closed in death

The Leeds express with a fearful crash
Rushed into her like lightning's flash
The dreadful cries were heard around
From the dying ones on the snowy ground
The cries for help did rend the air
From husbands, wives, and children dear
Who were cut down in a moment time
By that sad disaster upon the line

Poor Mr. Sanderson's grief we fear
For his two daughters is hard to bear
To think those poor girls are no more
Their untimely fate he does now deplore
No more his daughters in life he'll see
His wife's heart is broke with misery
For the loss of those whom they loved well
Their sorrow is more tongue can tell

In many a home there's a vacant chair
For some daughter or husband dear
Wives and mothers in grief now? Mourn
For the loss of those who are dead and gone
In heaven they hope they again will meet
With the King of Kings at the mercy seat?
Them prap(sic) for those beneath the clay
Who were killed that morn on the railway

The Abbots Ripton rail disaster occurred during a fierce blizzard on 21 January 1876 on the Great Northern Railway line. The Flying Scotsman, the *Special Scotch Express* train from Edinburgh to London, was involved in a collision with a coal train; the King's Cross to Leeds express 'at full speed in a flurry of flying snow' travelling in the other direction then ran into the wreckage.[31] 13 passengers died and 53 passengers and six train crew members were injured.

The initial accident was caused by over-reliance on signals when travelling at high speed in adverse conditions, coupled with systematic signal failure owing to accumulation of snow and ice. The signal 'automatically dropped back to "all clear" because of the weight of the frozen snow on the long signal wire'.[32] The subsequent inquiry into the accident 'had a permanent influence on railway practice[33] and led to fundamental changes in British railway signalling practice.

Broadside ballads began to decline in popularity towards the end of the nineteenth century when they could not keep up with other, newer, forms of cheap print resulting from technological innovations in printing. At the same time the invention of the telegraph transformed the way that journalists and newspapers conducted business leading to a demand for up-to-date accurate news rather than long rhyming ballads sung in the streets.

Endnotes

1. H J Dyos and D H Aldcroft. *British Transport. An Economic Survey from the Seventeenth Century to the Twentieth.* (Harmondsworth: Penguin, 1969), 135.
2. Dyos, *British Transport*, 129.
3. Dyos, *British Transport*, 155.
4. BBC. 'Broadside ballads: When the news was spread through song', September 16, 2016. https://www.bbc.co.uk/news/uk-england-37386026
5. Paxton Hehmeyer. 'The Social Function of the Broadside Ballad; or, a New Medley of Readers'. 2007 https://ebba.english.ucsb.edu/page/social-function-of-the-ballad
6. Ibid.
7. Manchester Library has several thousand ballad sheets dating from the seventeenth century to the end of the nineteenth century. The Bodleian Libraries also houses a digital collection of ballads.
8. Terry Coleman. *The Railway Navvies.* (London: Hutchinson, 1965): 26.
9. Helena Wojtczak. *Railwaywomen.* (Hastings, East Sussex: Hastings Press, 2005): 1.
10. Simon Bradley. *The Railways: Nation, Network and People.* (London: Profile Books, 2015): 337.
11. National Railway Museum. https://www.railwaymuseum.org.uk/research-and-archive/further-resources
12. The spelling varied until the 1870s.
13. Jeffrey Wells. *An Illustrated Historical Survey of the Railways in and around Bury.* (London: Challenger, 1995).
14. Coleman, *The Railway Navvies*: 28.
15. Bradley. *The Railways*: 337.
16. Ibid.
17. Coleman. *The Railway Navvies*: 105-6.
18. Coleman, *The Railway Navvies*: 22

19 Daniel O'Connell (1775 – 1847) was an Irish political leader and campaigner for Irish independence.
20 Carl Sandburg. *The American Songbag*. (New York: Harcourt Brace Jovanovich, 1927): 356.
21 British Library.'Poem by Wordsworth, Suggested by the Proposed Kendal and Windermere Railway'. http://www.bl.uk/onlinegallery/onlineex/kinggeorge/p/027add000044361u00278000.html
22 Jack Simmons. *The Victorian Railway*. (London: Thames & Hudson, 1991): 164.
23 Philip Bagwell and Peter Lyth. *Transport in Britain. From Canal Lock to Gridlock*. (London: Hambledon and London, 2002): 105.
24 By 1870 over 1,000 locomotives had been built by R. & W. Hawthorn.
25 https://spellerweb.net/rhindex/UKRH/NorthEastern/NewCarRly.html
26 Jack Simmons in 'Opening ceremonies' in The Oxford Companion to British Railway History. (Oxford: Oxford University Press, 1997): 357.
27 Jack Simmons in 'Newcastle & Carlisle Railway' in *The Oxford Companion to British Railway History*. (Oxford: Oxford University Press, 1997): 345
28 Alan Earnshaw. *The Holmfirth (Summer Wine) Branch Line* (Appleby-in-Westmorland, Cumbria: Trans-Pennine Publishing Ltd., 2005): 21-22 .
29 Office of Rail and Road blog. https://orr.gov.uk/news-and-blogs/orr-blog?result_23585_result_page=19
30 L T C Rolt. *Red for Danger. The Classic History of British Railway Disasters*. (Stroud: The History Press, 2009): 61
31 Rolt. *Red for Danger*, 87.
32 Ibid.
33 Rolt. *Red for Danger*, 88-9.

2 Music hall and the railway

MUSIC HALL WAS a booming industry in the UK during the late nineteenth century. By 1870, London had 33 music halls with an average audience capacity of 1500. Numbers of halls were growing in the provinces, Sheffield had ten and Manchester and Leeds both had eight,[1] and by this time most English towns had access to at least one. Musical life in general had received a boost from the huge growth in cheap rail travel: brass bands and choirs could travel the country to take part in competitions and concerts; and music hall artists could move easily from venue to venue, town to town, performing the same repertoire.[2] At the same time the music publishing industry was flourishing, largely because of the popularity of music hall. Towards the end of the nineteenth century most middle class homes owned an upright piano or a harmonium which became the focus of home entertainment: there was a huge market for sheet music with hit songs selling in their hundreds of thousands.[3]

Although the roots of music halls can be traced back in part to the rough drinking saloon concerts in the first half of the nineteenth century, the first true halls emerged in the 1850s and became increasingly up market. The first boom was in the 1860s and early 1870s with the halls becoming bigger, more expensive and luxurious in their décor and architectural design, until they eventually transformed into the more respectable variety theatres at the turn of the century. Going to the music hall was a social event: people went along in groups to enjoy the entertainment sitting at tables being served by waiters, eating dishes such as pease pudding, oysters and steak and kidney pie[4], drinking and smoking. The audience was part of the performance, heckling the performers and joining in with the choruses. Contrary to what is commonly supposed, the halls were not the exclusive preserve of the working classes.

Although a good proportion of the singers were working class, most of the composers were middle class and, in general, the demographic of the audience cut across classes with a mainstay of upper working class and lower middle class. Most were young men under 25, some were much younger than that, especially in the cheaper gallery seats.

Recent technological innovations in printing meant that music had become much easier to print. The system for printing scores had changed from using engraved metal plates to moveable type which was much more economic, as was the move to colour lithography for the pictorial covers. The frontispieces were often highly elaborate illustrations of contemporary scenes and music hall artists and are of great historic interest today. Music hall songs only required a couple of sheets of paper to produce and they sold well, consequently publishers of the time relied on this market.[5] However, the sheet music for music hall songs cost three or four shillings so most copies were found in middle class homes. All this meant that new songs could spread around the country quickly and become part of the national culture. Thousands of catchy new songs appeared dealing, often humorously, with matters of everyday life: love, marriage, current fashions, social comment and contemporary events. Songs were often written for specific artists, and them alone, to perform.

In an age when the railways had transformed so many people's lives, it is not surprising that there was a large repertoire of songs about railway journeys, station porters, guards, trains and the escapades of the passengers. As Rudyard Kipling wrote to his friend James Benion Booth 'Those old music-hall ditties…supply a gap in the national history, and people haven't yet realised how much they had to do with national life.'[6] In *Music and Politics*, John Street argues that music has the 'capacity to capture historical experience' and that it 'embodies or conveys politics'.[7] The lyrics of songs in the music hall looked at life from a working-class perspective – a social commentary on the way that ordinary people lived.

Many of the music hall songs told stories, some were sentimental, many were humorous often with an element of self-mockery, and countless rested on sexual innuendo and double entendre. In terms of the subject matter of songs connected with the railway, several common themes can be found: songs featuring railway employees; tales of railway excursions; and songs narrating the possibility of being tricked by a pretty girl when alone in a carriage, a preoccupation of contemporary Victorian society with cases of such being much reported in newspapers. The film critic John Huntley described music hall as the 'original permissive society' which recognised the 'utter stupidity of the artificially refined exterior with which Victorian society attempted to cover up its less savoury truths'.[8]

50 music hall songs about the railways

SONG TITLE	MUSIC HALL ARTIST	SONGWRITER(S)	DATE
Jobs on the railway			
The lost luggage man	George Robey		
Give me a ticket to heaven	Denham Harrison	Richard Elton & Denham Harrison	
All change for Llanfairfechan	Wilkie Bard	Worton David & George Arthurs	
I kept on waving my flag	George Formby Senior		
The Muddle Puddle porter	Lionel Brough	George Grossman Junior	
Railway porter, Dan	Henri Clark	Harry Hunter & G. D. Fox	1884
Oh! Mister Porter	Marie Lloyd		1892
The wheeltapper's song		Clifford Seyler & Wolseley Charles	1923
Johnny the engine driver	Annie Adams	G. W. Hunt	1867
The railway fireman	Harry Gordon		
The railway guard	Arthur Lloyd	Alfred Plumpton	1866
The railway guard	Will Fyffe	Will Fyffe	
The railway guard	Dan Leno	George LeBrunn	1890
The railway belle and the railway guard	Harry Clifton	Harry Clifton	1865
Excursion trains			
The cheap excursion train	Lillie Langtry	J. P. Harrington	1898
Rosie had a very rosy time	Marie Lloyd	Fred Murray	1909
I couldn't get in!	Charles Bignell	J. P. Harrington	
It ain't all honey and it ain't all jam	Vesta Victoria	Fred Murray & Geo. Everard	1906
The young man on the railway	Harry Clifton	W. H. Brinkworth	1865

A little idea of my own	George Robey		
Train journeys			
Show me a train that goes to London			
Don't stick it out like that	Fred Earle	Fred Murray & Fred W. Leigh	1901
I looked out of the window	Fred Earle		
Tickle me Timothy do	Billy Williams	R. P. Weston & F. J. Barnes	1908
He's gone where they don't play billiards	Sam Mayo	J. P. Harrington & Sam Mayo	
Let's go round and have a taster	Ernie Mayne		
The railway station sandwich	Will Evans	W. H. Wallis & F. E. Terry	1902
Dear old Shepherd's Bush	Nat D. Ayer	Nat D. Ayer & Clifford F. Grey	1916
And the leaves began to fall			
London, Chatham and Dover	"Jolly" John Mash	G. W. Hunt	1865
Take me back to dear old Blighty	Lily Morris, Ellie Retford		
The train that's taking you home	Will Fyffe	Will Fyffe	1929
I want you to notice my leggings	Wilkie Bard	F. Leo	
The underground railway	Marcus Wilkinson	Watkin Williams	1863
The Tuppenny Tube	R. G. Knowles	Edgar Bateman & Henry E Pether	1900
The husband's boat	Alfred Vance	Frank W. Green & Alfred Lee	1869
Bang went the chance of a lifetime	George Robey	George Robey & Sax Rohmer	1908
You can't punch my ticket again	Cissie Kent	Charles Collins	1897
Watching the trains come in	Jack Pleasants	Frank Leo	1916
Watching the trains go out	Jack Pleasants	William Hargreaves	1912
Oh blow the scenery on the railway	George Lashwood	F.W. Leigh & G. Arthur	1910
There's danger on the line. The great semaphore song	G. H. MacDermott	G. P. Norman	1875
Travelling in a confined carriage			
The kiss in the railway		G. H. Mackey	1864
Where do flies go in the winter time?	Jack Pleasants, Sam Mayo	Sam Mayo & Frank Leo	

Tommy make room for your uncle	Marie Lloyd, W. B. Fair, Tony Pastor	T. S. Lonsdale	1876
Oh! Mister what's-er-name	Marie Lloyd	Fred Murray & Kenneth Lyle	1907
Sexual symbolism of trains			
He missed his train again	Clarkson Rose		
I've never lost my last train, yet	Marie Lloyd	G. Rollit & George Le Brunn	1912
Lost her way	"Viscount" Walter Munroe		
What did she know about railways	Marie Lloyd	C. G. Cotes & Bennett Scott	1897

Most songs are in verse and chorus form with lilting melodies and catchy choruses. In musical terms, they were pretty straightforward because they had to be. The audience customarily joined in with the chorus, consequently many songs have a narrow melodic range making them easier to sing. The harmony is usually very basic with the use of a few simple chords, many using what is now referred to as the three chord trick. The most successful songs were the result of the combination of a good tune, witty memorable lyrics, and a good performer who often acted the part of the character they were singing about. 'Oh! Mister Porter', as sung by Marie Lloyd, one of the biggest music hall stars, combined all these factors and was hugely successful.

Marie Lloyd and Oh! Mister Porter

Marie Lloyd (1870 – 1922) was born in the working-class area of Hoxton in East London. She had a long career with a large number of hits, topping the bill at music halls in both the United Kingdom and the United States and acting in pantomime with two other music hall stars - Dan Leno and Tich. Her songs were frequently character songs, often quite risqué, and full of double meanings, one of the most famous being 'What did she know about railways' with the infamous lines *Someone wanted to punch her ticket/ The guards and porters came round by the score/ And she told them all she'd never had her ticket punched before.* This was during the period of middle-class moralism with its Victorian ideal of womanhood, one of wife and mother and the guardian of Christian virtues- a 'cult of domesticity which rested firmly on the double standard of sexual conduct.⁹ Peter Honri in *Working the Halls* writes that Marie Lloyd was 'one of the first people to recognise that there was a great deal to be said for 'doing and speaking in public what Victorian society did in private'.¹⁰ In some songs the trains themselves were symbolic of sexual activity. For example, in her song, 'I've never lost my last train, yet' the chorus uses the 'last train' to symbolise a 'modest little maiden's' virginity.

Yes, I've learnt to know the bliss
Of a stolen little kiss
When you heave a sigh and softly murmur "Pet"
As you gaze into his face
Wrapt in amorous embrace
But I've never lost my last train yet, Oh no
I have never lost my last train yet.¹¹

Oh! Mr Porter, Miss Marie Lloyd

Chronicle / Alamy Stock Photo

'Oh! Mister Porter' was written in 1892 by George Le Brunn (1864 – 1905) and his brother Thomas and was taken on tour that year by Marie Lloyd. It was a huge success. The song skips along in 6/8, moving mainly by step to make it easy for the audience to join in the chorus. As the writer Max Beerbohm commented after a performance 'The flurry, the frantic distress of it …Marie was bursting with rapture and made us partakers of it'.[12] On the face of it, 'Oh! Mr Porter' is about a young woman, an innocent abroad, returning from a visit to her old aunt in London and getting on the wrong train which goes beyond her stop. However, the story goes on to reveal that she meets an old gentleman who grasped her leg, pulled her back and she sank into his arms where she laid her trembling head. To quote the writer Arnold Bennett of a performance by Lloyd at the Tivoli Theatre: 'All her songs were variations on the same theme of sexual naughtiness. No censor would ever pass them, and especially he wouldn't pass her winks and silences.'[13] So when we bear in mind Lloyd's infamous nods and winks and telling looks, another reading might tell us that she 'has gone too far'. The last verse includes the lines *If you make a fuss of me and on me do not frown / You shall have my mansion, dear, away in London town.* The fantasy of being saved from poverty by a rich benefactor, often an older man, is another common theme in music hall songs.

Lloyd made a huge amount of money in her lifetime but she never forgot her modest roots. In an 1895 interview in the *Daily Sketch* it was reported that she was 'paying nightly for one hundred and fifty beds for the homeless and destitute of "Darker London".'[14] When the music hall workers went on strike in 1907, she gave generously to the strike fund and was there on the picket lines. At the age of 52 she collapsed on stage at the Edmonton Empire and died three days later. The writer Max Beerbohm described her funeral as the best-attended in London since that of the Duke of Wellington. The poet T S Eliot reflected on her death in the American literary magazine *The Dial*:

> whereas other comedians amuse their audiences as much and sometimes more than Marie Lloyd, no other comedian succeeded so well in giving expression to the life of that audience, in raising it to a kind of art.[15]

Oh! Mister Porter

Lately I just spent a week with my old Aunt Brown
Came up to see the wond'rous sights of famous London Town
Just a week I had of it, all round the place we'd roam
Wasn't I sorry on the day I had to go back home.
Worried about with packing, I arrived late at the station
Dropped my hat box in the mud, the things all fell about
Got my ticket, said good-bye "Right away" the guard did cry
But I found the train was wrong and shouted out,

Chorus: Oh! Mr Porter what shall I do
I want to go to Birmingham
And they're taking me on to Crewe
Send me back to London as quickly as you can
Oh! Mr Porter what a silly girl I am.
The porter would not stop the train.

But I laughed and said, "You must
Keep your hair on Mary Ann, and mind that you don't bust".
Some old gentleman inside declared that it was hard
Said, "Look out of the window, Miss and try and call the guard."
Didn't I, too, with all my might I nearly balanced over
But my old friend grasp'd my leg, and pulled me back again
Nearly fainting with the fright, I sank into his arms a sight
Went into hysterics but I cried in vain,

Chorus

On his clean old shirtfront then I laid my trembling head
"Do take it easy, rest awhile," the dear old chappie said
"If you make a fuss of me and on me do not frown
You shall have my mansion, dear, away in London Town."
Wouldn't you think me silly if I said I could not like him?
Really he seemed a nice old boy, so I replied this way
"I will be your own for life, your immy doodle um little wife
If you'll never tease me any more I say.

Chorus

The cheap excursion train

With the arrival of the railways came the first excursion trains – almost from the very start. Within a few weeks of its opening in 1830, the London and Midland Railway (LMR) took an excursion of 150 Manchester Sunday school children to Liverpool. Cheap excursion trains were hugely popular in Britain throughout the nineteenth century and into the twentieth century, much more so than anywhere else on the continent. It is easy to see why such train trips were instantly successful. Prior to this, journeys by road were on foot, horseback or in a carriage (something that was too expensive for most people). Excursions for pleasure usually took place on water, on canals or rivers, and sometimes in steamboats, but even steamboats did not have the capacity to carry large numbers.

Excursion trips by train, on the other hand, had reduced fares making them available to many people and having the capacity to transport huge numbers of them. The Great Exhibition of 1851 attracted thousands of visitors carried there by trains, no exact numbers are known because returns were not given, but they probably ran into the hundreds of thousands. Most trips were to the seaside, pleasure grounds, and sports event including prize fights (even though they were illegal). They were organised by newly created travel agents, business works, friendly societies, and the temperance societies which flourished at the time. Many trips were for pure pleasure but others, often run by the mechanics' institutes, were instructional. There was a seemingly endless stream of passengers keen to take the opportunity to travel outside their immediate locale and, in order to cater for them, railway companies added more and more carriages. In August 1844 an excursion was organised from Wakefield to Hull, with 97 carriages. This was a record for the time.

A MONSTER TRAIN

> A special pleasure train of the Manchester and Leeds Railway left Wakefield for Hull on Monday last, consisting of 97 carriages, with the extraordinary, and we believe, unprecedented number of 4000 passengers, composed of the "Ancient Order of Foresters", other lodges, and their friends.[16]

The music hall song 'The cheap excursion train' was written and composed by J. P. Harrington (1865 – 1939) in 1898.[17] It was written for Lillie Langtry (1877 – 1965) a music hall comedian known as the Electric Spark (not to be confused with her contemporary the well-known actress Lillie Langtry, the 'Jersey Lily'). By the time this song was written, monster trains were no longer around, instead several trains were put on at once when needed. Contemporary accounts of nineteenth century railway excursions depict them as jolly affairs, high-spirited, often chaotic and certainly memorable. Music and singing often played a part. Street's claim that music has the 'capacity to capture historical experience' is reinforced by the first verse and the chorus of 'The Cheap Excursion Train' where all these characteristics of excursions are captured in the lyrics.

The reference to 'Ladies smoking' is interesting. Smoking was generally prohibited on trains from the 1830s to 1868 when the Railway Regulation Act made it compulsory for railway companies to provide designated carriages for smokers, before this the approach had been inconsistent and people smoked 'in defiance of the bye-laws and the penalties' as evidenced by this passage from *The Railway Traveller's Handy Book of 1862*:

SMOKING

We believe that on certain occasions the stringency of the prohibition is relaxed, and we were once present at the arrival of a train from Doncaster, during race time, which train had more the odour of a divan18 than anything else, and every other passenger who alighted therefrom, held between his lips a partially consumed cigar.[19]

Although women were permitted to use smoking compartments, the carriages were seen as a male domain and women who encroached upon this territory were perceived as 'invaders'. Smoking was seen as the preserve of prostitutes and lower-class women: on the whole middle-class women did not smoke, those who did so were regarded as 'fast'.

The second verse gives us another snapshot of late nineteenth-century society in its depiction of the upper-class toff 'Bertie, with his eye glass' and his fancy way of speaking. The excursion trip was an important part of working-class culture, but the social composition of the passengers on the trains cut across classes. The 'swell', as such types with their resplendent dress and arrogant manner were commonly known, was a stereotype much parodied in the music hall to the extent that there was a particular type of music hall performer who specialised in their portrayal known as the *lion comique*. The well-known song 'Champagne Charlie' was amongst their repertoire.

Verse three encapsulates a common attitude held in the late nineteenth century in its description of the way that men took advantage of the darkness of the confined space of the carriage when travelling through a tunnel and used this as an opportunity to grope women (or worse). This concern was fuelled by regular and sensationalist reports of attacks and sexual assaults on female passengers in the press. *The Railway Traveller's Handy Book of 1862* provides the following cautionary advice:

CAUTION IN PASSING THROUGH TUNNELS

> Male passengers have sometimes been assaulted and robbed, and females insulted, in passing through tunnels…In going through a tunnel, therefore, it is always as well to have the hands and arms ready disposed for defence, so that in the event of an attack, the assailant may be instantly beaten back or restrained.[20]

The cheap excursion train

Puff, puff, puff, puff goes the engine, now we're starting off again
For a day out in the country by a cheap excursion train
Chaps and girls sit in the corners, all's as cosy as can be
Some have seats and others haven't, some are glad to take a knee
Doors a-banging - banjos twanging to the latest song
Ladies smoking - fellows joking - 'Now we shan't be long.'

Chorus: *Now we're off again, on a cheap excursion train*
Any amount of chaps and gals, all in their Sunday falderals
Singing a lively song, 'Life on the ocean main'
Going away for a lively day by the cheap excursion train.

Then there's Bertie, with his eye-glass
Thinks the company really is beastly, bally, awfully vulgar
With its songs and sandwiches, just imagine Bertie's horror
When a lady shouts out quick ' 'Ere I say - you with the winder
Hold my baby, 'arf a tick.'
Some reciting - others fighting, swearing all the way
Babies squalling - porters bawling- there's a peaceful day.

Chorus

Now we've got into a tunnel, all's as lively as can be
'Keep your hands out of my pockets.' 'Who's that tickling my knee?'
'You leave off, you, Bill! I know you- I can tell you by your squeeze'
'Leave off -do! Oh mother, mother. Porter turn the lights up please.'
Lights are gleaming - girls cease screaming- they once more look gay
Gay, but rumpled - hats all crumpled - what a tunnel, eh?

Chorus

It would be wrong to give the impression that all music hall songs are full of sexual innuendo. Another characteristic of music hall lyrics was their play upon words. 'Railway porter, Dan' is a good example of this with its humour resting on the witty use of puns on place names, for example: *If you're fond of billiards go to Kew and then to Poole / And to the Scilly Islands if you fancy you're a fool and The Milkmen should go more to Cowes, I very often say / And Jockeys should be off to Ryde, or else Horse-trail-i-a.* The song was written and composed by Harry Hunter and G D Fox for the performer Henri Clark.

Railway porter, Dan

I am a Railway Porter and my mates all call me Dan
I earn as many ha'pence as a Railway Porter can
My wages are a pound a week, which isn't very fat
When there's a wife and family to see to out of that
It takes me all my time I know, to find them bread and cheese
And I should have no dinners but for my gratuitous
Gratuities are sixpences and threepences and bobs
Which I get hold of now and then for waiting on the nobs

Chorus: *Take your seats for Leicester, Chester, Birmingham and Crewe*
See your luggage labelled, for the place you're going to
And if you are a Captain or a Colonel or a man
You can find a little sixpence for the Porter Dan

Now if you'd like a journey that will suit you, don't you know
Just tell me what you are and then I'll tell you where to go
If you're fond of billiards go to Kew and then to Poole
And to the Scilly Islands if you fancy you're a fool
If you're a waiter you should go of course to Table Bay
And if you are a donkey, then you ought to go to Bray
Go to the Sandwich Islands if you're hungry and would dine
But if you're fond of bloaters go to Herrin'-on-the-Rhine

Chorus

There's the Isle of Man for ladies who are looking out for mates
For lovers who've made up their minds, there's the United States
For those who want divorces, Sunderland of course is best
And all the babies certainly prefer to go to Brest
All those who like a game of whist should cut at once for Deal
And pugilists to Box Moor go when they have been to Peel
Bookbinders should be happy, if to Russia they are bound
And invalids be glad if they arrive at Plymouth Sound

Chorus

The Milkmen should go more to Cowes, I very often say
And Jockeys should be off to Ryde, or else Horse-trail-i-a
The Mashers should to Starch Green go to keep their collars stiff
And Tenor singers make a note to go to Tenerife
The girls should go to Kissingen, there is no doubt they do
The boys should go to Darlington, to Ems and Nancy too
The Queen should go to Queensland though it's far across the sea
And the Prince of Wales in Kingsland then would very welcome be.

Chorus

Porters were hired by railway companies from the beginning of the railways with LMR being the first to employ them in 1830. Their general duties were to give directions to board trains, to stow passengers' luggage and then to unload it and to carry it to the awaiting omnibuses. Porters in large stations operated in gangs with specific tasks whereas in small country stations porters also took on more general tasks from ticket selling and collecting, to cleaning waiting rooms. Many country porters also took on the role of signal man. Porters were often recruited from the country.[21] Physical standards were high, employees needed to be healthy, have good eyesight and, for the majority of companies, should be able to read and write.[22] For most of the nineteenth century, they were paid 17 shillings a week, which, although it was low, was an improvement on, for example, the wages of an agricultural labourer. Some railway companies prohibited porters from accepting gratuities, but in practice tips were usually accepted in order to supplement low pay. *The Railway Traveller's Handy Book of 1862* gives the following advice:

HIRING PORTER

If a person does not require a vehicle, he will seldom have any difficulty in finding a porter to carry his luggage. A crowd of men and boys are usually to be met with outside the terminus who, for a trifling fee, will convey the luggage to any part of the town. It would be well to hire a clean and honest-looking messenger, and also to take the precaution of making him walk on in front, as you will then have your eye upon him, and he cannot well decamp without your knowledge. On the principle of seeking information from everybody, you may, while jogging onwards, glean from your luggage-bearer such items of intelligence as will prove advantageous.[23]

The Music Hall Artistes Railway Association (MHARA)

Music hall artistes were constantly on tour and, although the growth of rail travel meant that they were able to move from town to town, this was not without both expense and difficulties. There were often huge amounts of equipment to be transported: not only the artistes and their personal luggage, but also scenery, props, and costumes. Against a background of general disquiet with the railways, music hall artistes were constantly seeking improvements to both fares and timetabling. During the late nineteenth century they formed pressure groups, notably the MHARA, and negotiated with the railways to secure better travelling conditions for their number.

The Railway Regulation Act 1844 and the Cheap Trains Act of 1883 Acts had addressed some passenger issues, ensuring that 3rd class passengers should be protected from the weather at a fare of 1d per mile and that these fares should be exempt from taxes, but the acts only set out the minimum quality obligation of rail passenger service. Improvements beyond this were still left to the railway companies.[24] At a time when railways provided the only reliable means of long-distance travel, a large part of the problem for theatre companies was that the weekly change over from one theatre to another, ready to work again on Mondays, had to take place on Saturday nights because of the slowness and infrequency of trains on Sundays. This was very inconvenient. In the early 1880s John Bosworth, an enterprising railway clerk employed by Midland Railway, came up with a better system to accommodate the needs of theatrical traffic. Bosworth organised special Sunday trains and subsequently became a manager of theatrical traffic for Midland Railway. Bosworth's system involved theatrical companies booking their journeys in advance which meant that he was able to schedule his Sunday Specials in such a way that one train could cater for several theatre companies at once. There are several letters and articles of fulsome praise for the Sunday Specials in *The Era*, then known as 'The Great Theatrical Journal'. On the 12 December, 1885, it included the following report:

> The Midland Railway conveyed thirty theatrical and operatic companies over their line on Sunday and Monday last, numbering 613 members of the profession. As many as 252 members arranged to travel by one train on Sunday… from Liverpool to London. Fifty seven vehicles were provided for the convenience of the parties. Fifty two of these belonged to the Midland Company, and the arrangements were efficiently carried out by Mr John Bosworth, the theatrical travelling agent of that company.[25]

However many more pieces in *The Era* during the 1880s reflected a general disquiet with railway companies and their price discrimination, an object of suspicion and led many to perceive them as monopolists'.[26] This ability to discriminate caused problems for theatre companies in the constantly changing charges the railway companies made for luggage. The following extract from a short article is typical. It was written at a time when changes were made to the amount of excessive baggage that could be carried on the train without incurring charges.

> ...What confidence will be felt by capitalists in starting touring enterprises under a sort of Reign of (Railway) Terror, when none can say what the next despotic decree may be? Will not the railway companies find themselves the losers, by the constant diminution of theatrical travellers? A stingy policy is generally a losing one in the end. Let the Boards think over the matter, and perhaps they may rescind or modify their resolution...[27]

Later that year *The Stage* made the point that 'The amount paid in fares by the profession to the railway companies must form a very valuable addition to their receipts, a little courtesy and attention should therefore be paid in return'.[28]

Partly as a result of this dissatisfaction with the railways, the Music Hall Artistes Railway Association (MHARA) was formed in 1897 and by 1900 had a list of 5,196 members.[29] Its founding members included several leading artistes: Dan Leno;[30] the ventriloquist F W Millis; and the comedian Charles Coborn. The MHARA was officially apolitical and focused on campaigning for railway travel concessions.[31] In 1897 *The Stage* reported that 'a deputation from the MHARA' approached the Railway Clearing House (RCH) 'on a question of a reduction of railway rates'.[32] Consequently, it negotiated special train fares for the transport of touring companies and their baggage. Members received a 25% discount on train fares when travelling in parties of five or more, considerably reducing the expensive costs. Several more meetings with the RCH followed in the quest for further concessions.[33] These meetings are well documented by contemporary theatre papers such as *The Stage* and *The Era*.[34]

Membership of the MHARA continued to rise and in 1905 *The Stage* reported the 'committee wishes to heartily thank the several theatrical managers of the railway companies for their invaluable co-operation'.[35] In 1912 they met again with the RCH pursuing 'a peaceful path, securing for its privileged members a satisfactory reduction in travelling expenses'.[36] By way of example, a company of 30 travelling 50 miles with two trucks to carry all of their equipment would pay 75% of the standard fare rate for each member of the company. They would get one baggage truck free and would pay for the other at 3d/mile.

This is not to say that all the travelling problems for music hall artistes were over. In 1906 the MHARA joined up with other workers' organisations to form a trade union – the Variety Artistes Federation and in 1907, boosted by the membership of the musicians and stage hands unions, they took part in a music hall strike against working conditions. These conditions included the banning of last minute changes in venues and timetables, a matter of much concern for the large number of artistes who often performed in several theatres in one week and needed to book in advance planning to get from venue to venue as cheaply and efficiently as possible.

In 1914, following a large number of fires in music halls, London County Council imposed stricter controls which included banning eating and drinking in the auditorium – part of the essence of music hall entertainment. World War I saw a surge in popularity as performers were rallied to promote the war with patriotic calls to arms and morale-boosting songs. However, the interwar years saw a decline of the halls and a more respectable type of music hall entertainment arrived under the name of 'variety'. Competition from cinema, upcoming jazz and swing bands, and the outlet of radio led to the gradual demise of music hall and, by the onset of World War II, music hall had virtually disappeared.

Endnotes

1. http://www.bl.uk/onlinegallery/onlineex/vicpopmus/t/015hzz00001772du00031006.html
2. Dave Russell. *Popular music in England. 1840-1914: A social history* (Manchester: Manchester University Press, 1987).
3. Ruth Towse. "Copyright Auctions and the Asset Value of a Copyright Work." *Review of Economic Research on Copyright Issues* 13 no. 2: 83-99.
4. http://www.bl.uk/onlinegallery/onlineex/vicpopmus/t/015hzz00001772du00031006.html
5. Towse, "Copyright Auctions", 85.
6. P J Keating. *The Working Class in Victorian Fiction*. (London: Routledge and Kegan, 1971): 160.
7. John Street. *Music and politics*. (Cambridge: Polity Press, 2012).
8. John Huntley cited in Peter Honri's *Working the Halls*. (London: Futura Publications, 1974): 89.
9. J Mitchell and A Oakley. *The Rights and Wrongs of Women*. (London: Pelican, 1976): 61.
10. Honri, *Working the Halls*, 91.
11. Chorus of 'I've never lost my last train yet', written and composed by G. Rollit & George Le Brunn, 1912.
12. Richard Anthony Baker. *British Music Hall: An Illustrated History*.(Barnsley: Pen & Sword Books Ltd, 2014): 138
13. Arnold Bennett. *The Journals of Arnold Bennett 1896-1910*. (London: Casell Ltd., 1932).
14. *Daily Sketch*, 25 December, 1895.

15 'London Letter' in December 1922 edition of *The Dial*.
16 *The Scotsman*, 14 August, 1844
17 As advertised in *The Era*, 16 July, 1898.
18 The word 'divan' is used here to mean a coffee house or smoking room.
19 Jack Simmons, ed. *Railway Traveller's Handy Book of 1862*. (Bath: Adams & Dart, 1971): 66-7.
20 Simmons, *Railway Traveller's Handy Book*, 74-5
21 Simmons, *Railway Traveller's Handy Book*, 98-9.
22 P W Kingsford. *Victorian Railwaymen. The emergence and growth of railway labour 1830-1870*. (London: Frank Cass, 1970): 9.
23 Simmons, *Railway Traveller's Handy Book*, 98-9.
24 Huroki Shin. "Have consumer movements enhanced transport justice? Passenger representation on Britain's railways before 1947." In *Transport policy: learning lessons from history*. Colin Divall, Julian Hine, Colin Pooley eds. (Burlington, VT: Ashgate Publishing Company, 2016): 81-3.
25 "Theatrical Gossip" in *Era*, 12 December, 1885.
26 P J Cain, "Railways and price discrimination: the case of Agriculture, 1880-1914". (*Business History* 18 no. 2, 1976): 190.
27 "The Profession and the Railways" in *The Era*, 12 April, 1884.
28 *The Stage*, 21 November, 1884.
29 *The Era*, Dec 1, 1900.
30 Famous for singing the music hall song 'The Railway Guard'.
31 Peter Bailey, ed. *Music Hall: The Business of Pleasure*. (Milton Keynes: Open University Press, 1986): 180.
32 *The Stage*, 17 June, 1897.
33 Railway Clearing House Committee Minutes and Reports. https://discovery.nationalarchives.gov.uk/details/r/C1793093
34 Other reports of the activities of the MHARA can be found in the *Music Hall and Theatre Review, The Referee and the London and Provincial Entr'acte*.
35 *The Stage*, 19 October, 1905.
36 L Carson, ed. *The 1913 Stage Yearbook* (London: Carson & Comerford, 2015): 135.

3 Railway works bands, choirs and musical societies

Railway brass bands

Railways and brass bands have long been associated. The first association, and one familiar to many, is the long tradition of brass bands playing at railway stations to mark significant occasions, arrivals and departures. This started in the nineteenth–century when it became customary to play Handel's 'See the conquering hero comes' to celebrate the opening of a line or station, bands also often accompanied early railway excursions. By the end of the nineteenth century there were more than 7000 brass bands in the UK[1] and audiences for open-air concerts and contests were measured in the tens of thousands.[2] Without the railways this would not have been possible: the bands and their considerable fan base would not have been able to get to the national championships, which were often some distance away. They were also events which some players considered to be the *raison d'être* of the movement. This is not the only association between the bands and the railways. The Victorian establishment 'held music and, in particular, art music to be a force for moral and positive good among working people'.[3] and, from the early days of the movement, brass bands were often linked to a work force. So too were choirs and, to a lesser extent, orchestras and bell ringers.

In 1847 *The Musical Times* wrote of the Great Western Railway (Swindon) Band

> In the large workshops of the Great Western Railway, at Swindon, a number of these men have combined to make a most excellent orchestra, seconded by the liberality and encouragement which seems to pervade the Company's arrangement at this village, for the benefit, improvement and amusement of their workmen.

The brass band movement was particularly strong in the North of England, bands were often linked to workplaces including railway companies and works. The largest category is that of the colliery bands followed by those related to the railways.[4] Two such bands were the Leeds Railway Foundry Band and the Doncaster Loco Band. Their stories tell us much about both nineteenth-century brass bands and railway history.

The Leeds Railway Foundry Band

The Railway Foundry was a railway engineering workshop in the Hunslet area of Leeds. It was established in 1838 by Shepherd and Todd. Orders were soon received from the Liverpool and Manchester Railway, the Leeds and Selby Railway, Hudson's York and North Midland along with a couple of locomotives which were exported to the Paris & Orleans Railway. In 1844 there were 40 members of staff, by 1847 there were 400. This was the year that the company produced their most famous engine, the *Jenny Lind*. It was built for the London, Brighton and South Coast Railway and became the basis of one of the company's most successful standard designs. By now 'railway mania' was in full swing and the Railway Foundry's use of standard designs meant that new engines could be bought off the shelf. Many engines were based on the *Jenny Lind* and these were exported all over the world. In 1858 the Leeds Foundry had produced over 600 engines and by 1860 it had become the biggest railway workshop in the world. It went on to build engines for around another 100 years.

The Leeds Railway Foundry Band started its life in 1851 as the Fairbairn's Wellington Saxhorn band but soon changed its name when it became attached to the foundry.[5] The 1850s was a significant period in the development of brass bands. The bands' means of funding fell largely into three types: the first was those, such as the Leeds Railway Foundry band, who were linked to a single workplace or were the beneficiaries of wealthy patronage; the second type was subscription bands who relied for their support from the wider community such as mechanics institutes or temperance societies or indeed the members themselves.[6] The third type belonged to the burgeoning 1859 volunteer movement. Many works bands were also partly funded by subscriptions. The thinking behind such industrial patronage stemmed from Victorian paternalistic ideology in the belief that it helped to 'improve' the workers.[7] There were further advantages for bands that were linked to a workplace: membership gave improved job security; as time went on, it became common for outstanding players to hold retainers for their services; and there was more flexibility to allow players to miss work to perform in concerts and contests. Works bands also provided entertainment for their fellow workers. The repertoire was very largely arrangements of operatic works and soon thousands of people were introduced to art music through the brass band. At the same time winning contests became a matter of local pride and importance for the workforce.

The first large-scale brass band contest took place at Belle Vue, Manchester in 1853. For many years (1836-1970), there was a zoo and amusement park at Belle Vue Zoological Gardens as well as an annual circus: it was also a popular destination for excursions in the early days of the railways. The first excursion train to Belle Vue ran on the Manchester and Birmingham railway - from Macclesfield to Longsight station. The growing rail network and the provision of cheap excursion fares soon meant that contests became the setting for bandsmen drawn from wide geographical areas. Although only eight bands took part in the first contest, *The Manchester Guardian* estimated that there were 16,000 people in attendance and rather snootily reported that

> ...all the bands, without exception chose selections from operas and concerted pieces, the majority of which were intricate and difficult of execution indeed such as might have taxed the highest powers of professional performers. And when it is considered that all the members, without exception, (although upwards of one hundred), belonged to the working classes, and studied music, not professionally but for recreation, the performances were in the highest degree creditable both to their taste and proficiency.[8]

Although the Leeds Railway Foundry Band did not take part in the first Belle Vue contest they did enter the second one in September 1854 and went on to win the first prize of £20. The contest attracted 14 bands, with 191 performers in all, and an even larger audience of 20,000. Most of the spectators were brought by special trains from West Yorkshire, East Lancashire and Staffordshire. Some of the audience would have been Belle Vue's normal pleasure seekers but the *Manchester Guardian* reported that a large proportion of those present were there for the music, 'proud of the efficiency of the bands from their respective localities'. The report continues,

> All the bands were what are called country bands, consisting, not of professional musicians, whose whole time is devoted to the study and practice of the art, but of hard working artisans, who have found its almost unaided study and practice an intellectual and elevating pursuit during the intervals between labour and repose.

The Leeds Railway Foundry was conducted by Richard Smith (1820-1890) who became an important figure in the early Belle Vue contests. Smith acted as both leader and conductor of the band which meant that as well as conducting he played the solo cornet part, a practice common at the time and not given up until near the end of the century. In addition to the cash prize won by the band, Smith received a bombardon worth 10 guineas.[9] The award of special prizes of instruments was to become an important feature of major contests when bands were always in needed of more, often expensive instruments. Following the announcement of the results the winning band returned to the platform and played Handel's 'See the conquering hero comes'. A year later Richard Smith took second place with the same band, and in 1856 he took the top two places: Leeds Railway Foundry came first and Leeds (Smith's) came second. It seems that this was a controversial decision. Shortly after the contest, members of Leeds (Smith's) band complained bitterly of this decision and put up a 'Musical Challenge for £100' to the Railway Foundry Band. They wrote:

> The decision at the recent contest at Manchester, certainly struck every one who heard the bands play on that occasion as the most barefaced and ridiculous ever sent out to the public; and that gentlemen possessing the situations and standing the judges are supposed to have, could be hoodwinked to give it, excited the disgust of the public, amateurs, and musicians assembled on the occasion, who received it with mingled hisses, cheers, and groans...WE HEREBY CHALLENGE THE RAILWAY FOUNDRY BAND, who received the Prize Instrument, to play for a purse of equal stakes of £50, £100 or any higher sum, either band to and as they now stand, the leaders singly, or instrument against instrument singly, in one week from this date, the money to be staked in the hands of some disinterested and respectable party; and thus expose one of the worst pieces of jugglery and imposition ever practised on the public at a pretended brass band contest; the greatest part of the persons assembled declaring that they (Railway Foundry Band) played the worst but one on the occasion. We have received offers from other bands, and from private individuals, to back up to any amount and expressive of their disapprobation, and believe that it was all agreed before the contest took place in consideration of CERTAIN WEIGHTY MATTERS, which no doubt will be brought to light afterwards.

An immediate reply, addressed to Smith's Leeds Brass Band, 65, Stamford Street, Leeds, will have prompt attention. From the Members of SMITH'S LEEDS BRASS BAND.

In the six years from 1853 to 1858, 34 different bands competed at Belle Vue, seven of which were works bands, four of these railway bands – as well as the Leeds Railway Foundry band were foundry bands from Bradford, Doncaster, and Tidswell's Foundry Band (Leeds). As can be seen, all these bands were from Yorkshire 'confirming the domination of that county at that time'.

The Doncaster Loco Band

The Great Northern Railway established the Doncaster Locomotive and Carriage Building Works in the early 1850s, a plant works which later became famous for building the LNER locomotives *Mallard* and the *Flying Scotsman*, as well as many thousands more locomotives.

The Doncaster Loco Band formed in June 1852, £50 having been spent on instruments. The band did not have a regular bandmaster or musical director and it was short lived. However, in 1856 a new band was formed: the Doncaster Plant Works Band (sometimes known as the Great Northern Railway Plant Works Band). George Birkinshaw, the father of the celebrated nineteenth-century Black Dyke cornet player of the same name, was the conductor. Its instrumentation included several brass instruments which had been popular in the early days of brass bands - six cornopeans, two ophicleides, and an Eb bass bombardon - but were mostly dropped in the 1860s and replaced by more modern instruments.[10] The band won several prizes between 1859 and 1861. In September 1859 they won the first prize of £15 at Peterborough here the bandsmen had to 'defend themselves physically against disappointed rivals afterwards'. They were renamed the Doncaster Volunteer Band in 1859, giving many outdoor concerts in the town often playing arrangements of marches, polkas, gallops, quadrilles and waltzes composed or arranged by their conductor Birkinshaw. The works band lasted until around 1910 and was followed by Doncaster National Union Railway Band which was active around the late 1920s and early 1930s.[11]

The Crystal Palace National Brass Band Contest and the role of the railways

One of the first national brass band championships took place at Crystal Palace in 1860. After the Great Exhibition in 1851, which had attracted thousands of visitors many on special excursion trains, the re-constructed Crystal Palace opened in 1854 in the South London suburbs in a newly enlarged form. Two stations were needed, both were originally at the ends of branch lines and exceptionally large for a South London suburb. One is now defunct. The former railway engineer George Grove was the company secretary and he wished to host orchestral concerts and oratorio festivals. These were highly successful, Grove wrote the programme notes, eventually retiring from his Crystal Palace post to work for the publishers MacMillan where he edited the famous Grove Dictionary of Music and Musicians.

The Crystal Palace National Brass Band Contest was organised by Enderby Jackson, a talented musician and entrepreneur from Hull. In collaboration with railway companies, Jackson secured 'inexpensive railway day excursions for bandsmen and their supporters…from the north to the contest' organising excursion trains that would 'run from all principal towns on the Great Western, Midland, London and North Western, Great Northern, South Western, and other railways'. All bands paid the same ticket prices regardless of the distance they had to travel to Sydenham.[12] Jackson paid the railway companies up front and then covered his costs with the entry fees the bands had paid.[13] Posters proclaimed -

> National Brass Band Contest - Crystal Palace, Sydenham
> **CRYSTAL PALACE - GREAT NATIONAL BRASS BAND CONTEST**
> **TUESDAY, 10th, and WEDNESDAY, 11th July.**
>
> Arrangements have been made for a GRAND MONSTER BRASS BAND CONTEST on the above days, in which upwards of ONE HUNDRED BANDS from all parts of England are engaged to take part, comprising in all TWO THOUSAND BRASS INSTRUMENT PERFORMERS. Valuable Prizes, in Money and Cups, will be given by the Company; and, in addition, the principal Music's' instrument makers In London have signified their intention to present several first-class instruments as special prizes.
>
> The Contest will commence each day at, Ten o'clock, and on both days the whole of the Bands will meet at Three o'clock precisely in the Handel Orchestra, and perform Mendelssohn's "Wedding March", Haydn's Chorus, "The Heavens are Telling", Handel's "Hallelujah", "Rule Britannia," and "God Save the queen." A Monster Gong Drum, seven feet in diameter, manufactured expressly for the occasion, will accompany the combined bands.
>
> Admission, Tuesday, Half-a-Crown; Wednesday, One Shilling.
>
> NOTICE.-Excursion Trains will run from all principal towns on the Great Western, Midland, London and North Western, Great Northern, South Western, and other railways; for full particulars of which see the Companies' advertisements and bills.

The Times estimated that on the second day of the 1860 Crystal Palace concert the audience 'being considerably over 22,000'.[14] The 44 bands performed in turn before the appointed judges, on platforms erected in various parts of the building. They each had to play two pieces, one of their own choice plus a set of quadrilles composed by Enderby Jackson. The most popular choices were arrangements of operatic arias by Bellini, Donizetti, Rossini and Verdi. Next they assembled together to play some of the staple pieces of brass band repertoire of the time - 'Rule Britannia', 'The Hallelujah Chorus', Mendelssohn's 'Wedding March', Haydn's 'The heavens are telling' from The Creation and finishing the massed performance with 'God save the Queen'. *The Times* reported

> The effect of the combined legions of "blowers" (upwards of 1,200 strong) was tremendous. The organ, which accompanied them, and which on less exceptional occasions is apt to drown everything, was scarcely audible in the midst of the brazen tempest. Nothing less than the new "monster gong-drum," manufactured by Mr. Henry Distin[15] - to wield the thunder of which required the united efforts of Messrs. Charles Thompson (of the Crystal Palace Band) and Middleditch (of the London Rifle Brigade)-could prevail against

it. The pieces that pleased the most-perhaps because the best executed-were Mendelssohn's Wedding March and the National Anthem, both of which were unanimously encored. The whole performance was conducted, with wonderful vigour and precision, by Mr. Enderby Jackson, of Hull.[16]

Of a later contest in 1865, the *Musical World* reported that 'each of the bands is accompanied by the hundreds of admirers from the surrounding country for miles around' with 'an immense influx of excursionists early trains' playing to a huge audience of 60,000.[17] Brass band contests were important partly because they offered material rewards in the form of prize money and instruments,[18] but also because they acted as the principal forum for the movement. The North soon emerged as the most successful area for bands.[19] Hence the contests not only engendered a feeling of pride and community exemplifying working class achievement and potential but also led to 'a spectacular rise in performing standards by brass bands, at a pace unequalled at any other time in any other type of music-making'. [20]

The number of brass bands increased until the 1890s and then started a slow decline. This decline had steepened by 1914 when many bandsmen signed up to join the armed forces. Numbers recovered on their return but fell again in World War II and from that point on there were significantly fewer bands in comparison with the earlier glory days. One railway band in particular stands out during the inter-war years – the Horwich Railway Mechanics Institute Band. The band formed in 1899 as the Lancashire and Yorkshire Railway Company Band, three years after the opening of the new locomotive works in Horwich. The intention was to recruit from railway employees but only one member of the old Lancashire and Yorkshire Band passed the audition. Consequently other members were recruited from Black Dyke, Irwell Springs and Foden's and in 1915 the band took second place at Belle Vue. After a strike at Foden's motor works, a dozen players who had been pickets were sacked and promptly moved to the Horwich band. The band's greatest achievement was to win the National Championship of Great Britain at Crystal Palace on 23rd September 1922, becoming 'Champions of Great Britain and the Colonies' and receiving the 1000 guineas trophy. The *Bolton Journal* reported their triumphant return home.

> As the time for the bands arrival from London approached, the streets became alive with people, and soon dense crowds had gathered outside the station and

lined the approach to the platform. It appeared that the whole of Horwich had turned out en-masse to give the champion band a rousing reception…The train…was gaily decorated with bunting and evergreens, and on the front of the engine was the device 'welcome home.' Loud cheers were given, fog signals were exploded on the line, and The Horwich Old Band played a lively air as the train steamed into the station. Headed by Horwich Old Band, and loco workers carrying torch lights, the band drove in a charabanc passing cheering spectators…cries of 'Bravo Horwich' could be heard above the din.[21]

It is difficult to make an accurate count of the number of railway works bands, partly because of the their transitory nature and changing names, but the brass band researcher Gavin Holman in *The works brass band – a historical directory* has included the following bands connected to railway companies, locomotive works, the National Union of Railwaymen and railway foundries.[22] The following list does not claim to be comprehensive, but it provides a useful overview of the number of bands connected in some way to the railways.

50 railway brass bands

BAND NAME	DATES ACTIVE
Belfast and Northern Counties Railway Brass Band	Active in 1872
Blyth LNER Band - Northumberland	1921 -late 1950s
Bradford Railway Foundry Band	
Bristol and District NUR Band	1920s – 1930s
Cambridge Railway Band.	1920s – 1930s.
Cardiff GWR Workers Band	1901 – 1930s
Cricklewood Midland Locomotive Brass Band	1890s
Crewe Locomotive Steam Works Band, sometimes known as the *Crewe Steam Sheds Band*. Its leader was manager of the steam sheds at the railway works.	1863 – 1930s
Darlington LNER Band	
Doncaster Railway Foundry Band	1850s -1860s
Dubs Loco Works Brass Band – Glasgow. Dubs and Company was a locomotive works founded in 1863. In 1903 it became part of the North British Locomotive Company which built 4,485 locomotives.	1900 – 1906
Edlington NUR Brass Band – Doncaster, Yorkshire	
Epping Town Band Founded by a Great Eastern Railway engine driver. Many original bandsmen were railway workers and the first bandmaster another railway worker.	1894 – early 1920s
Gateshead LNER Band	Active in 1932
Glasgow Loco Works Band Also known as *North British Locomotive Company Band*	1900S – 1920s.
Gorton Foundry Band Based at Beyer Peacock Ltd manufacturer of locomotive engines. Around 8,000 engines were built there between 1855 and 1966.	1880s
Great Eastern Brass Band – Bristol	1870s-1880s
GER Band – Ipswich, Suffolk	1860s
GER Works Band – Stratford, Essex	Late 1850s – 1920s
GWR and Metropolitan Band – London	1950s
GWR (Swindon) Band – Wiltshire	1847 - 1900s
Great Western Paddington Brass Band Also known as *Great Western Railway Carriage Works Brass Band*	Active in 1856 and 1865
GWR Brass Band – Worcester	Active in 1888
GWR Employees Brass Band – Taunton, Somerset	Active in 1895
GWR Social, Education and Union Association Band – Wiltshire	1918 – 1947
Hastings Southern Railway Band – Sussex	Active in 1928

Hawthorn's Locomotive Works Brass Band – Newcastle upon Tyne. R.W. Hawthorn & Co's Locomotive works.	Active in 1867
Horwich Lancashire and Yorkshire Railway Company Band Formed three years after the opening of the new locomotive works.	1889 - 1830
Leeds Railway Foundry Band	
Manchester and Sheffield Railway Company Locomotive Band	1860s
Midland Railway Loco Works Band Also known as *St Pancras Midland Railway Band*.	1890s and 1900s
Newton Abbot Locomotive Brass Band South Devon Locomotive Works.	1855 – 1863
Newcastle LNER Band – Northumberland	1930s
New Cross Locomotive Band Had several names including *New Cross Railway Mission Brass Band*.	
London, Brighton and South Coast Railway Locomotive Department Brass Band – New Cross	1890s and 1900s
New Holland NUR Silver Band – Lincolnshire	1890s – 1920s.
Newton Heath Loco Works Band – Lancashire	1950s
Newcastle Locomotive Works Band Also known as *Stephenson's Operatic Brass Band*. Based at the Robert Stephenson & Co. Locomotive works.	Early 1860s – mid-1870s
North Staffordshire Railway Locomotive Band	
Rotherham Locomotive Brass Band	Active in 1868
Reading Foundry Brass Band Great Western Railway Foundry	1860 still active in 1875
Reading NUR Band	Active in 1921
Peterborough Great Northern Railway Locomotive Brass Band	1857 – early 1860s
Sheaf Works Prize Band – Lincoln, associated with Ruston, Proctor & Co. factory which produced locomotives and steam shovels.	1890s to the 1920s
South Devon Locomotive Works Brass Band – Newton Abbot. South Devon Locomotive Works.	1855, still active in 1863
Southern Railway Works Brass Band – Lancing, Sussex	
Swindon British Railways Social and Educational Union Silver Band	1948
Tidswell's Foundry Band	
Willesden Junction LMS Railway Steam Shed (Silver) Prize Band	Active in 1907
Wolverhampton Railway Works Band also known as the *GWR Institute Band*. The GWR Institute - Stafford Road Railway Works.	1858 – 1970s

Abbreviations

GER – Great Eastern Railway

GWR - Great Western Railway

LMS - London, Midland and Scottish Railway

LNER – London and North Eastern Railway

NUR – National Union of Railwaymen

Railway choirs and musical societies

From the first half of the nineteenth century large choirs were established throughout the country, by 1860 there existed a Great Northern Railway Glee and Madrigal Society based at King's Cross, and towards the end of the century, male voice choirs started to become fashionable. Some of the factors which had led to the rapid increase in the number of brass bands in the nineteenth century - the industrial revolution, Victorian paternalism and philanthropy, an increase in leisure time and an appetite for public entertainment - also led to the huge expansion in the number of choirs and choral societies. At the same time music festivals sprang up all over the country and the railways played a large part in their success.

The choral movement shared the same competitive element as the brass band movement. The British Federation of Musical Competition Festivals was formed in the early 1920s and again the railways helped to facilitate the competitions. The Federation negotiated Special Railway Vouchers with the Railway Clearing House enabling choirs to travel reasonably cheaply. Some choirs were accommodated in special excursion trains. In December 1922 Mexborough and Swinton (Yorkshire) Railwaymen's Male Voice Choir organized its first competition, with an entry of twelve choirs. For the large number of choral societies, oratorios, particularly Handel's *Messiah*, were part of the stable repertoire. A common feature of such festivals was a massed performance on the final day and in 1857 at one of the most famous, the Crystal Palace triennial Handel Festival, *Messiah* was performed with an orchestra of 500 and a chorus of 2,000.

In the early twentieth century many of the railway choirs came under the auspices of Railway Musical Societies.

The Great Eastern Railway Musical Society

The Great Eastern Railway Musical Society was created in 1908. It comprised an orchestra and choir whose members were drawn entirely from railway staff. William Johnson Galloway, a director of the Great Eastern Railway

Company, was largely responsible for its creation and sometimes took part himself as the conductor. Their performances were of a high standard, enough to attract guest conductors of a high calibre such as Sir Henry Wood and Sir Alexander Mackenzie and to have compositions written for them, for example *Men on the Line*, a cantata for men's voices by Hubert Bath, composer of the more well-known *Cornish Rhapsody*.

Several of their concerts took place in the Hamilton Hall, Abercorn Rooms of the Liverpool Street Hotel. Many of these concerts were described as 'smoking concerts', informal concerts where audiences were grouped around tables, refreshments were served between the items and the audience were free to smoke, along with, at times, the performers themselves. Smoking concerts first became popular towards the end of the nineteenth century, as described in *The Musical Times* of February 1882:

> The recent establishment of Smoking Concerts in the metropolis is scarcely so much a proof of the advance of smoking as of the advance of music. The fact is that many persons accustomed to enjoy a cigar or pipe in the evening, and also exceedingly fond of listening to the performance of good works, have begun to see that the gratification of the one desire need not interfere with the occasional gratification of the other, and the result is the growth of the entertainments at one of which a few evenings ago we 'assisted'. Of course with a full orchestra, and a programme containing some of the best of our standard compositions, not only the total absence of ladies, but the arrangement of tables intermingled with seats, appeared strange to one accustomed to attend evening concerts at St. James's Hall; but then the stiffness inseparable from fashionable assemblies was replaced by an air of luxurious enjoyment which appeared thoroughly in consonance with the feelings of the audience: and when the performance commenced the few who desired to converse were effectually hushed by the frowning looks of the musical majority. We can confidently affirm that the characteristic feature of the concert was faithfully preserved, for not only the audience smoked, but the Conductor, the stringed instrument players, and the performers upon wind instruments too when they could get a chance. It was remarked by many that Beethoven sounded much better when, instead of sitting between two elegantly dressed ladies in a sofa stall, you could recline at your ease, a and combine the aroma of the music with the fragrance of the weed.[23]

Performances were not limited to Hamilton Hall, the musicians would travel to venues such as Queen's Hall and to cities outside London including

York and Edinburgh. The Science Museum holds a collection of concert programmes which include details of concerts given to HM Forces in the YMCA Hall, White City (1915), delegates to the International Railway Congress (1925) and a commemoration concert to mark the retirement of a Works Manager (1932).[24]

In 1923 Great Eastern Railway was grouped with other railways to form the London & North Eastern Railway (LNER). Consequently the society's name changed to the London & North Eastern Musical Society in 1924.

The London and North Eastern Railway Musical Society

The LNER covered Eastern England and Scotland and during the 1930s it had several male voice choirs scattered throughout this area based at Doncaster, Peterborough, Selby and elsewhere. The Doncaster LNER Musical Society had an orchestra as well as a male voice choir. This was the idea of their conductor H A Bennett who directed them in the late 1920s when he was organist at Doncaster Parish Church.[25] Each year the societies combined for a performance in London under their musical director the English choral conductor Leslie Woodgate (1900-1961). Woodgate joined the BBC in the 1920s and in 1934 he was appointed the BBC Chorus Master. He became one of the foremost choral trainers in the country: for several years he conducted Huddersfield Choral Society and he appeared at the Henry Wood Proms in 1946 conducting the BBC Singers. Woodgate composed music for the LNER, as did Stanley Marchant (1883-1949) organist at St Paul's Cathedral, and railwayman C F Chudleigh Candish, who wrote the popular *Song of the Jolly Roger* for male voices.[26]

There was a gradual decline in the number of bands between 1895 and 1914, numbers recovered slightly after World War 1 when soldiers returned and communities started to rebuild. The number of bands fell again after World War II but did not recover in the same way.[27]

At the time of writing few bands and choirs with railway connections remain.

10 current railway brass bands

BAND NAME	DATES ACTIVE
York Railway Institute Band Formed by merger of the York City Brass Band and the Ebor Excelsior Silver Band. York Railway Institute Golden Rail Band.	Founded in 1952
City of Exeter Railway Band Former names: Exeter & District Southern Railways Band, Exeter British Railways (Southern Region) Band, Exeter British Rail Band.	Founded in 1945
March Brass 2000 (Cambridgeshire) Former names: March Railway Servants Brass Band, GER Brass Band, March Railway Silver Band.	Founded in 1891
Netherfield Railway Band (Nottinghamshire) In 1950 was renamed Carlton Silver Band, then Nottingham City Transport Band in 1988. The band spawned Carlton Brass (a junior section) which went on to become the current Carlton Brass Band.	Founded in the 1890s
South West Trains Woodfalls Band (Wiltshire) Former names: Woodfalls Temperance Band, South of England Temperance Band, Woodfalls Silver Band, SWT Woodfalls Silver Band.	Founded in 1874
Swindon Pegasus Brass (Wiltshire) Former names: Swindon (Gorse Hill) Brass Band, Great Western Railway Social, Education and Union Association Band, Swindon Rail Staff Association Band, British Railways Silver Band (Swindon), Swindon BR Band, Swindon Concert Band.	Founded in 1912
Tylorstown Brass Band (Glamorgan) Former names: Tylorstown & District Silver Band, Tylorstown & Mardy Colliery Band, Tylorstown (Valley Lines) Brass Band, Tylorstown Band Arriva Trains Wales.	Founded in 1920
Watford Band (Hertfordshire) Former names: London and North Western Railway Band, L&NW Artizan Staff Brass Band, Watford Artisan Silver Band, Watford Silver Prize Band, Watford Silver Prize and British Legion Band, Watford United Ex-Service Men's Band, Watford Silver Band, Hosier & Dickinson Watford Band.	Founded in 1895
Wolverton Town Band (Buckinghamshire) Former names: Wolverton Town Band, Wolverton British Railway and Town Silver Band (1951-1989).	Founded in 1908
Tilbury Band (Essex) Former names: Tilbury Railwaymen's Band, Tilbury Town Silver Prize Band, Tilbury Town Band.	Founded in 1919

Finally there is one band that after well over 150 years is alive and well and is still heard by many, playing a weekly concert on the concourse of Paddington station. The Great Western Railway Paddington Band started its life in 1855 as the Great Western Railway Paddington Station Military Band (brass and woodwind) when it played Christmas music outside the stations master's cottage. It has remained active in some form ever since. In 1910, under the name Great Western Railway and Paddington Borough Prize Silver Band, it took part in the National Brass Band Championships at the Crystal Palace coming 8th out of 15 bands. In the mid-1920s it reverted to its military band format, a format which it remains in today.

Endnotes

1. Gavin Holman. *How Many Brass Bands? – An analysis of the distribution of bands in Britain and Ireland over the last 200 years*, 2018.
2. Gavin Holman has identified nearly 20,000 distinct brass bands in the UK between 1800 and the present.
3. Trevor Herbert. *Bands: The Brass Band Movement in the 19th and 20th Centuries* (Milton Keynes: Open University Press, 1991): 9.
4. Gavin Holman. *The works brass band – a historical directory of the industrial and corporate patronage and sponsorship of brass bands.* 2020.
5. Saxhorns were a family of brass instruments with valves developed by the Paris inventor of the saxophone – Adolphe Sax. Their popularity was short lived, but some saxhorn instruments live on in brass bands being similar to tenor horns, baritones and euphoniums.
6. The bandsmen of Epping Town Band, for example, each paid an entry fee of one shilling and subscriptions of threepence a week towards purchase of music. Money earned by the band was either divided among the bandsmen or put towards uniform purchase.
7. Herbert, *Bands*, 23.
8. *The Manchester Guardian*, September 10, 1853.
9. A bombardon is a bass instrument with valves, similar to a tuba and popular during this period.
10. Ophicleides were later replaced by valve basses and contrabasses, for example.
11. Gavin Holman. *Brass bands of the British Isles – a historical directory.* 2018.
12. Trevor Herbert and Arnold Myers. "Music for the multitude: accounts of brass bands entering Enderby Jackson's Crystal Palace contests in the 1860s." *Early Music* 38, no. 4 (2010): 572-84.
13. Ibid
14. *The Times*, July 12, 1860.
15. Henry Distin (1819 – 1903) was a cornet player and brass instrument manufacturer; born in England but working in America. He improved the design of brass valve instruments and took out several patents on new percussion instruments.
16. *The Times*, July 11, 1860.
17. Stephen Etheridge. "Southern Pennine Brass Bands and the Creation of Northern Identity c. 1840–1914: Musical Constructions of Space, Place And Region." *Northern History* 54, no. 2 (2017): 250.
18. Herbert, *Bands*, 37.
19. Etheridge, "Southern Pennine Brass Bands", 246.

20 Herbert, *Bands*, 7.
21 Stephen Etheridge. *Brass Band Contests and Railway Travel: Mobility, Audience Support and Sporting Comparisons*. 2017 https://bandsupper.wordpress.com/author/gtrombone/
22 Gavin Holman. *The works brass band – a historical directory of the industrial and corporate patronage and sponsorship of brass bands*. 2020.
23 *The Musical Times*, February, 1882.
24 https://collection.sciencemuseumgroup.org.uk/people/ap233/london-north-eastern-musical-society
25 Philip Scowcroft. "Railways in Music". http://www.musicweb-international.com/railways_in_music
26 Ibid.
27 Holman, *How many brass bands?* 2018

4 Gilbert and Sullivan and the railway

OF COURSE MUSIC hall was not the only form of music theatre in the nineteenth century. The operettas of Gilbert and Sullivan were hugely popular, not only in the Savoy Theatre but across the country at a time when amateur operatic societies were springing up all over the UK.[1] The producer Richard D'Oyly Carte brought together the dramatist W S Gilbert (1836-1911) and the composer Arthur Sullivan (1842-1900) and nurtured their collaboration. Together they created 14 operettas, the best known being *The Mikado, The Pirates of Penzance* and *HMS Pinafore*. In 1881 D'Oyly Carte built the Savoy Theatre and founded the D'Oyly Carte Opera Company specifically to perform their works. The operettas embraced social and political satire by the means of ridiculous plots; the words were satirical and the tunes were catchy.

The railways were something close to Gilbert's heart. In 1897 he wrote to *The Times* to complain about the lateness of the trains on the line between Watford and Euston during the holiday season.[2] The letter famously ends with the words 'To the question, "What has caused the train to be so late?" the officials reply, "It is Saturday" — as who should say, "It is an earthquake."'[3]

Gilbert is similarly critical in three of the references he makes to the railways in his operettas. 'The Lord Chancellor's nightmare song' from *Iolanthe* (1882) includes a 'very small second class carriage' as a feature of the nightmare. Second class carriages during the 1880s would have been quite uncomfortable; the lighting would have been poor and very few carriages would have been heated before 1900. It should be noted, however, that by 1882 few railway companies offered second class travel – this intermediate class had been largely abandoned leaving a choice between the expensive first class and the cheap, but overcrowded, third-class carriages.

For you dream you are crossing the channel, and tossing
About in a steamer from Harwich,
Which is something between a large bathing machine
And a very small second class carriage,

And you're giving a treat (penny ice and cold meat)
To a party of friends and relations,
They're a ravenous horde, and they all come aboard
At Sloane Square and South Kensington stations.

A further critical reference to carriage classes is made in Trial by Jury in 'The judge's song'.

In Westminster Hall I danced a dance,
Like a semi-despondent fury;
For I thought I never should hit on a chance
Of addressing a British jury –
But I soon got tired of third-class journeys,
And dinners of bread and water;
So I fell in love with a rich attorney's
Elderly, ugly daughter.

In the 1885 operetta *The Mikado*, the Mikado's diatribe which opens with the words 'A more humane Mikado never' goes on to prescribe punishments to fit various crimes. The fourth verse announces that:

The idiot who, in railway carriages,
Scribbles on window-panes,
We only suffer
To ride on a buffer
In Parliamentary trains.

Parliamentary trains originated in 1841 following a terrible train accident at Sonning, near Reading which resulted in the deaths of eight people who were travelling in low-sided open wagons. This led to a Board of Trade enquiry into the conveyance of third-class passengers. As a result Gladstone's 1844 Railway Regulation Act ensured 'the provision of at least one train a day each way at a speed of not less than 12 miles an hour including stops, which were to be made at all stations, and of carriages protected from the weather and provided with seats; for all which luxuries not more than a penny a mile might be charged.' In return the railway companies became exempt of the 5% passenger tax which was normally levied on all fares. In 1844 the minimum speed of 12 miles per hour was relatively fast, even the best horse-drawn carriages could go no faster than 10 mph. However, by the 1870s, the minimum speed of Parliamentary trains, as they came to be known, was considered to be very slow. Nevertheless, they stimulated the growth of cheap travel and enabled thousands of people to travel who otherwise would not have been able to travel at all. Parliamentary tickets did not disappear for a long time, Bishop's Castle Railway was still issuing tickets marked 'Parliament' in 1935.

The Junction Song

'The Junction Song' (sometimes known as 'The Railway Song'), 'a cautionary song about an egalitarian railway chairman'[4] is full of references to the railway and had an accompaniment including the sounds of 'a railway bell, a railway whistle, and some new instrument of music imitating the agreeable sound of a train in motion.'[5] It is sad to say that the music was never published and it is now lost so performances today either adapt music from other Sullivan scores or use music specially written by more recent composers. The song was sung by the title character of the first work that Gilbert and Sullivan collaborated on - *Thespis, or The Gods Grown Old* - an operatic extravaganza that was premièred in London at the Gaiety Theatre on 26 December 1871. So successful was the Christmas entertainment that it ran into early March the following year. It was advertised as 'An entirely original Grotesque Opera in Two Acts' and included 'much dancing and magnificent scenery and costumes'.[6] It tells the story of the aging Gods on Mount Olympus who decide to leave for a holiday.

We know from contemporary reviews that the orchestration included a railway bell and whistle as well as the sound of a train in motion. In its review, the stage magazine *The Era* also describes the 'ballad…being admirably sung by Mr. J. L. Toole, and furnished with a screaming, whistling, and shouting horns [which] fairly brings the house down'.[7] 'The entire company join in the chorus, the music of which admirably expresses the whirl and thunder of a railway train at express speed.'[8]

'Twas told to me with great compunction,
By one who had discharged with unction
A chairman of directors function
On the North South East West Diddlesex Junction.
Fol diddle, lol diddle, lol lol lay.

The chorus emphasises the rhythmic train-evoking sounds of the rhyming words 'function', 'junction', 'compunction' and 'unction' ending with the rhythmic 'Fol diddle, lol diddle, lol lol lay'.

This could suggest the following mechanical rhythm.

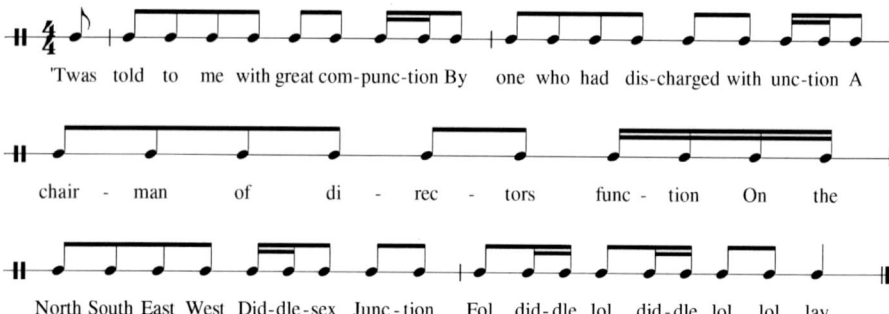

The Junction Song from Thespis.

I once knew a chap who discharged a function
On the North South East West Diddlesex Junction.
He was conspicuous exceeding,
For his affable ways, and his easy breeding.
Although a chairman of directors,
He was hand in glove with the ticket inspectors.
He tipped the guards with brand new fivers,
And sang little songs to the engine drivers.

Chorus

'Twas told to me with great compunction,
By one who had discharged with unction
A chairman of directors function
On the North South East West Diddlesex Junction.
Fol diddle, lol diddle, lol lol lay.

Each Christmas day he gave each stoker
A silver shovel and a golden poker.
He'd button hole flowers for the ticket sorters
And rich Bath-buns for the outside porters.
He'd moun the clerks on his first-class hunters,
And he build little villas for the road-side shunters,
And if any were fond of pigeon shooting,
He'd ask them down to his place at Tooting.

Chorus.

In course of time there spread a rumour
That he did all this from a sense of humour.
So instead of signalling and stoking,
They gave themselves up to a course of joking.
Whenever they knew that he was riding,
They shunted his train on a lonely siding,
Or stopped all night in the middle of a tunnel,
On the plea that the boiler was a-coming through the funnel.

Chorus

If he wished to go to Perth or Stirling,
His train through several counties whirling,
Would set him down in a fit of larking,
At four a.m. in the wilds of Barking.
This pleased his whim and seemed to strike it,
But the general public did not like it.
The receipts fell, after a few repeatings,
And he got it hot at the annual meetings.

Chorus

He followed out his whim with vigour,
The shares went down to a nominal figure.
These are the sad results proceeding
From his affable ways and his easy breeding.
The line, with its rails and guards and peelers,
Was sold for a song to marine store dealers
The shareholders are all in the work'us,
And he sells pipe-lights in the Regent Circus.

Chorus

Each of the verses, barring the first, can be fitted to a similar rhythmic pattern using repeated quavers and pairs of semiquavers. The first verse does not fit into this metric pattern probably because it was not intended to be sung, but was delivered in a declamatory *parlando* (speaking) style.

Several writers have taken up the idea that the song is about the Duke of Sutherland, referring to him as the Director of the London North Western Railway (LNWR), who had a fondness for trains and reputedly enjoyed riding along on the footplate of his company's engines.⁹ The Duke of Sutherland in 1871, the time that *Thespis* was written, was George Granville William Sutherland-Leveson-Gower, 3rd Duke of Sutherland (1828 –1892). He did indeed have a great fondness for trains and in 1870 he financed the extension of the Highland Railway line from Golspie to Helmsdale via his own home - Dunrobin Castle. In return he received a private station for his castle and the freedom to run his own train on the lines owned by the Highland Railway, having had his own carriages and locomotive built for the purpose. The Duke's enthusiasm for railway building projects and steam technology was well-documented, but this went together with a 'decided lack of interest in politics, a traditional duty of the aristocracy'.¹⁰ In its 1892 obituary of the Duke, *The Times* reported that 'He was an active director of the London and North-Western, of the Highland, and other railways…and he was ever ready to assist in the development of ingenious ideas in machinery, mechanical appliances, and the like.'¹¹ To some he was seen as a dabbling plutocrat. The satirical magazine *Punch* wrote that 'he is clearly the wrong man in the wrong place on any platform except that of a steam engine'.¹²

In 1895, his son, the 4th Duke of Sutherland ordered a new locomotive and carriages naming the locomotive Dunrobin, after his castle. One of the unique features of the locomotive was the enlarged footplate and enclosed cab with a four-person upholstered seat set high up at the back. This may be where the idea of the 3rd Duke of Sutherland enjoying riding on the footplates of trains came from. Those riding in its cab were to include King Edward VII, King George V, King George VI, King Alfonso of Spain, Kaiser Wilhelm II and Sir Winston Churchill. ¹³

Endnotes

1. The National Operatic and Dramatic Association, in 1899, reported that in 1914 nearly 200 British amateur operatic societies were producing Gilbert and Sullivan works that year.
2. For the last two decades of his life W S Gilbert lived at Grim's Dyke, his house in North West London. The nearest station was Harrow and Wealdstone, a stop on the Watford to Euston line.
3. *The Times*, September 28, 1897.
4. Philip Scowcroft. "Railways in Music". http://www.musicweb-international.com/railways_in_music
5. *The Era*, December 31, 1871.
6. *Pall Mall Gazette*, January 3, 1872.
7. *The Era*, December 31, 1871.
8. *Sunday Times* cited in S Tillett and R Spencer. *Forty years of Thespis scholarship*. 2002. https://www.gsarchive.net/thespis/Thespis40.pdf
9. Carolyn Williams. *Gilbert and Sullivan. Gender, Genre and Parody*. (New York: Columbia University Press, 2011): 52.
10. Annie Tindley. *Sutherland Estate, 1850-1920. Aristocratic Decline, Estate Management and Land Reform*. (Edinburgh: Edinburgh University Press, 2010): 51.
11. *The Times*, September 24, 1892.
12. *Punch*, January 26, 1878.
13. National Railway Museum. https://blog.railwaymuseum.org.uk/dunrobin-royal-engine/

The coming of the railways to Europe

THE FIRST RAILWAYS were constructed in Great Britain carrying passengers and providing the much needed transportation of natural resources, such as coal and iron, to fuel the Industrial Revolution. After the early success of the Liverpool and Manchester line in 1830, several European countries followed suit with the development of the railways hastening the pace of industrialization across Europe. For both freight and passengers, the railways proved to be a better and cheaper alternative than waterways and roads, many of which were impassable during the winter. The railways had an effect on many aspects of society: they stimulated economic development; they were a significant tool in nation building; and for some countries they were an important instrument of military strategy. The impact of the railways was huge in terms of progress and development. The railways created new towns, new industries and new concepts; railways became a symbol of modernity. The historian Eric Hobsbawm wrote

> By 1850, the railways had reached a standard of performance not seriously improved upon until the abandonment of steam in the mid-twentieth century, their organization and methods were on a scale unparalleled in any other industry, their use of novel and science-based technology (such as the electric telegraph) unprecedented. They appeared to be several generations ahead of the rest of the economy, and indeed 'railway' became a sort of synonym for ultra-modernity in the 1840s, as 'atomic' was to be after the Second World War.[1]

Thousands of kilometres of lines had transformed time, space and distance. The speed of even the earliest locomotives outstripped any means of transport that had come before, so given the enormity of the impact and the far-reaching effects, it comes as no surprise that the railways appealed to the artistic imagination of nineteenth-century composers. Jeffrey Richards and John MacKenzie when writing of the impact of the railways on popular culture put this appeal as 'responding to the immediacy of the sensations they provoke, to the bold iconographic power, strength and modernity of the steam engine, to the kaleidoscopic nature of train travel, to the whirl and bustle and breathlessness of transience'.[2]

Train Landscape, Ravilious, Eric, 1903–1942
Photo credit: Aberdeen Art Gallery & Museums

5 The coming of the railways to Austria, the Strauss family and railway music

THE FIRST PUBLIC railway in Austria was also the first in continental Europe, a horse-drawn line with an unusual gauge; it opened in 1832 and ran from Budweis[3] to Linz, eventually developing into a network of lines stretching for more than 270 kilometres in the valleys of Upper Austria. However, the first line in Austria to become part of the modern network was a steam hauled, standard gauge line, opened in 1837 between Floridsdorf (near Vienna) and Deutsch Wagram. This was the first section of the Kaiser Ferdinands-Nordbahn which eventually ran between Vienna and Warsaw. Austria's primary railway network developed rapidly in the mid-nineteenth century, partly in a bid to knit the Austro-Hungarian Empire together. By 1845 more than 650,000 passengers each year travelled along the Nordbahn line.[4] The Nordbahn followed a relatively simple route travelling largely across lowlands. In contrast the Südbahn, reaching south across the Alps to the Adriatic via the Semmering Pass, was built across difficult mountainous terrain which was more challenging in its construction with many tunnels and impressive multiple-arch viaducts. The Brenner Pass offered further difficult terrain, crossing avalanche runs with steep climbs, tight curves and long tunnels. In 1856 construction began westwards from Vienna with the Kaiserin Elisabeth Bahn, the first section of what was to become the Westbahn running from Vienna to Salzburg. It opened for traffic in 1860. Around this time railway construction more or less came to a standstill because of the war against Prussia but in 1866 it was pushed forward to revive the economy and the following years were marked by a railway construction boom.[5]

Railway building continued across Europe in phases. The last European country to build a railway was Greece; its first line, the Piraeus-Athens service, opened in 1869. By the early 1900s there was a grid of lines across Europe and several cross-border connections had been made creating a new and exciting network of technology.

There are many nineteenth-century pieces celebrating the coming of the railways; some were commissioned for the opening of lines; many evoke the scene using the instruments of the orchestra to imitate train sounds; and others focus on the danger and exhilaration associated with exciting new means of travel. For fifty years the Strauss family were at the forefront of the fashion for railway music.

The Strauss family

Johann Strauss I spent much of the early years of his musical career playing in pubs and he played a big part in establishing the waltz as a musical form, taking it to new heights of popularity. Robert Barry writes of Johann Strauss I that 'his career began playing the violin in the back room of a pub' and that he 'transmuted all the crowd-pleasing turns necessary for a bar-room entertainer into the new dance form of the Viennese waltz.'[6] Johann Strauss I (or the Elder as he is sometimes known) was born in 1804, the son of an innkeeper and was brought up in a Viennese tavern. He 'grew up amid the sounds of the wandering "Beer fiddlers" and the various folk dances they brought with them from their travels along the Danube.'[7] As a teenager he joined Joseph Lanner's string ensemble playing and writing dance pieces for the small group which eventually expanded to split into two popular string orchestras; Lanner conducted one and Strauss the other. In 1825 Strauss began giving concerts at *Zum roten Hahn* (The Red Rooster) where he met his wife to be Maria Anna. The couple married in July 1825 and their son, the future 'Waltz King', Johann II was born in October that year. They went on to have five more children; two of these Josef (1827–70) and Eduard (1835–1916) also went on to have successful careers as composers and conductors.

Strauss was energetic, prolific and very hard working. In 1829 his career was boosted when he signed a six-year contract with the owner of the fashionable *Zum Sperlbauer*, often known as the Sperl, the biggest dance hall in Vienna. In his early career Strauss appeared at dance establishments throughout Vienna, but realized that in order to secure an international reputation he needed to go further afield and soon began tours across Europe including two highly successful visits to Great Britain. In October 1838, Strauss set off on the lengthiest and most ambitious of his tours, first through southern Germany, on to Strasbourg and then Paris where the 26-strong orchestra stayed for four months. There his concerts received an outstanding reception and were attended by many prominent musicians including Cherubini, Paganini and Berlioz. The French composer Hector Berlioz wrote

> We knew the name of Strauss, thanks to the music publishers…that was all; of the technical perfection, of the fire, the intelligence and the rhythmic feeling which his orchestra displays, we had no notion.[8]

>   Journal des Débats (10 November, 1837)

Most of the touring would be by means of uncomfortable horse drawn carriages, but as the orchestra left Paris and continued through Belgium and the Netherlands they were able to take advantage of the new and extensive railway system. Largely as a result of Strauss's tours, the waltz gained huge popularity abroad, both in the ballroom and at promenade concerts such as those held at the Vauxhall Pavilion at Pavlosk near St Petersburg, and the Tivoli Gardens in Copenhagen. Periods of tours and residences in Vienna followed. In 1849, shortly after the second British visit with 46 concerts in eight weeks, Strauss returned to Vienna exhausted. He fell ill and a few weeks later he died from scarlet fever.

After his father's death, Johann Strauss II took over what had become the family business leading it to new heights of international fame. By the 1850s the Strauss business had 200 employees including coach drivers, music copyists and bookkeepers. In 1853 he broke down from exhaustion and handed over control to his brothers. To help his recovery he went to the spa town of Bad Gastein. Whilst he was there he was to take up another lucrative business opportunity when the Russian director of the Tsarskoe Selo Railway invited him to take up a summer residency at the station and entertainment venue in Pavlovsk near St Petersburg. The younger Strauss toured widely across Europe using the railway network that connected the German speaking spa towns. The spa resorts were an important market for the orchestra with their luxury hotels, casinos, theatres, restaurants and other popular entertainment venues for light music.[9]

The Strauss name was now synonymous with the waltz, even more so with the later appearance on the scene of Johann's two brothers, Josef in 1853 and

Eduard in 1859. Each of the three brothers had a stint of conducting the Strauss Orchestra. Although Josef had embarked on a successful career as an engineer and designer, he joined the orchestra in the 1850s and took over as conductor. In June 1870, Josef collapsed whilst conducting and fell unconscious. Seven weeks later he died at the young age of 42. After his untimely death, the youngest brother Eduard took over responsibility for the Strauss Orchestra.

In the 1890s Eduard Strauss undertook a highly successful tour of North America, along with concert tours through Europe and residencies in Vienna. In 1897 he discovered to his horror that his two sons, with his wife's knowledge, had squandered much of his earnings. Undeterred he set about recouping his fortune with another tour of North America. Whilst there he was injured in a train accident in Pittsburgh station; his collarbone was dislocated and he was forced to conduct with his left hand. The concert tour ended in New York in 1901 at which point he decided that after 39 years with the orchestra he had had enough. He disbanded the orchestra the following morning and retired from public life.[10] Six years later he took the Strauss Orchestra archives to a furnace in Vienna and burnt the lot. In 1916 he died of a heart attack and, as the Strauss expert, Peter Kemp wrote 'A golden epoch of Viennese musical history had been brought to a close'.[11]

Between them the Strauss family played a large part in creating the growing demand for light and accessible entertainment music, consequently introducing the popular-music industry to Europe.[12] In the words of Robert Barry, their music was 'unabashedly commercial; catering to an urban, middlebrow, leisure-time audience; and serving no religious or secular ritual function; Strauss's music represented a whole new way of looking at music'.[13]

The music of the Strauss family

10 railway pieces by members of the Strauss family

TITLE	COMPOSER	DATE
Eisenbahn-Lust-Walzer (Railway Pleasure Waltzes) Op. 89	Johann Strauss I 1804-49	1836
Spiralen (Spirals) Op. 209	Johann Strauss II 1825-99	1858
Accellerationen (Accelerations) Op. 234	Johann Strauss II	1860
Gruss an München (Greetings to Munich) Op. 90	Josef Strauss 1827-70	1860
Vergnügungszug (Pleasure Train Polka) Op. 281	Johann Strauss II	1864
Bahn Frei! (Line Clear) Op. 45	Eduard Strauss 1835-1916	1869
Mit Dampf (With Steam)	Eduard Strauss	1871
Tour und Retour (Round Trip)	Eduard Strauss	1875
Lustfahrten (Pleasure Journeys) Op. 177	Eduard Strauss	1879
Feuerfunken (Sparks of Fire) Op. 185	Eduard Strauss	1880

The waltzes

Waltz music is in triple time (three beats in a bar), quite fast, often with catchy tune and a simple oom-cha-cha chordal accompaniment. The dance with its fast-moving, twirling and embracing couples, at first shocked polite society because it required bodily contact between the dancers unlike the group dances and stately minuets which had come before. But it was this intimate sensuality which added to its popularity. The sensuous nature of dancing the waltz is vividly portrayed by Gustave Flaubert in this famously erotic description of a waltz in his novel *Madame Bovary*

> They began slowly, then went more rapidly. They turned; all around them was turning – the lamps, the furniture, the wainscoting, the floor, like a disc on a pivot. On passing near the doors the bottom of Emma's dress caught against his trousers.

> Their legs commingled; he looked down at her; she raised her eyes to his. A torpor seized her; she stopped. They started again, and with a more rapid movement; the Viscount, dragging her along disappeared with her to the end

of the gallery, where panting, she almost fell, and for a moment rested her head upon his breast. And then, still turning, but more slowly, he guided her back to her seat. She leaned back against a wall and covered her eyes with her hands.[14]

Gustave Flaubert: Madame Bovary

The second half of the nineteenth century was a time when the waltz craze was at its height. In 1833 the German journalist and critic Heinrich Laube wrote of an evening of dancing to Strauss's music at the Sperl, the venue which the Strauss Orchestra packed on a regular basis.

> To hold the unrestrained crowds in check a long rope is taken, and all who remain in the middle are separated from those actually occupied in dancing. But the boundary is flexible and yielding, and only in the steadily whirling girls' heads can one distinguish the stream of dancers. The couples waltz intoxicated through all the accidental or intentional obstructions and wild delight is let loose...[15]

Eisenbahn-Lust-Walzer (Railway Pleasure Waltzes)

The *Eisenbahn-Lust-Walzer* (Railway Pleasure Waltzes) were composed in 1836 in advance of the much anticipated 1837 opening of the line between Floridsdorf and Deutsch Wagram, the first section of the Kaiser Ferdinands-Nordbahn. The first performance took place on July 18, 1836 at the newly built dance hall which was part of *Zur golden Birn* (The Golden Pear), a very popular hotel and restaurant in the suburbs of Vienna.[16] The work was part of the venue's summer gala, an 'Assemblée mit Ball' (Assembly with Ball) under the title 'Buntes aus der Zeit' (Topical miscellany), performed at midnight against a 30-metre backdrop depicting a railway journey. The music was so rapturously received that it had to be repeated several times.

In the 1830s most waltzes, like the *Eisenbahn-Lust-Walzer*, were in sets of five plus the Introduction and a Coda (end section). The train journey gets off to a start in the Introduction with loud repeated brassy chords and tremolo strings leading up to dissonant off-beat accented chords. The whole arresting passage is repeated and then the excitement builds up more. After a silent bar the music suddenly becomes quieter, only to build up again with a climbing chromatic scale moving towards the final loud chords of the Introduction.

The five waltzes bear all the typical characteristics of the form: they are in triple time; quite fast; melodious; and with a simple chordal accompaniment. Most of the time the music uses an oom-cha-cha accompaniment where the bass note is played on the first beat of the bar and the remaining notes of the chord are played on beats two and three – oom-cha-cha, oom-cha-cha – a rhythm which lends itself to the sound of a train chugging along. Julian Johnson refers to this style of melody and accompaniment as a kind of 'proto-techno music which…operates with the thrum of the machine underlying a slow-moving melodic line'.[17] In Waltz No. 1, the train is on its way, chugging along in the way he describes:

The evocation of the train is also heard in the French horn sounds that are used to represent the sound of the train's whistle in the opening bar of Waltz No. 2:

The final section, the Coda, brings back the earlier waltz themes and the journey reaches its end with an emphatic cadence.

By this time Strauss's reputation had spread abroad and at the beginning of 1837, the *Eisenbahn-Lust-Walzer* were published, and well received, in England. There were several newspaper reviews including this one from the Brighton Gazette.

THE RAILROAD WALTZES by JOHANN STRAUSS

These are some of the very prettiest waltzes of this celebrated composer, and, in the present railway mania, are likely to become extremely popular, were it only for their name. They are, however, independently of that circumstance, extremely good.

Brighton Gazette (9 February, 1837)

Accellerationen (Accelerations)

It is interesting to note that although so many of the Strauss brothers' pieces are related to travel, none of them ever undertook journeys willingly and Johann II was apparently 'filled with terror at the mere mention of the precipices on the mountain stretches over the Semmering'.[18] Johnson writes that the 'new energy and kinds of motion engendered by new technologies' were often reflected in the titles of the Strauss pieces e.g. Johann Strauss II's 'Spirals' and 'Accelerations'.[19] He argues that dance titles such as *Ohne Bremse* (Without Brakes), *Feuerfunken* (Sparks of Fire) and *Ohne Aufenthalt* (Non Stop), 'play on the frisson of danger of the express train ride but also blur the distinction between the physical sensations linked to new technologies of transport with those of sexual excitement'.[20] Sigmund Freud was later to express this more directly when he wrote 'A compulsive link of this kind between railway-travel and sexuality is clearly derived from the pleasurable character of the sensations of movement'.[21]

This link between sexuality and the motion of a train is clearly portrayed in *Accellerationen* (Accelerations).[22] Written in 1860 for the Engineering Students' Ball held on February 14 in the Sofienbad-Saal and completed at the last moment, 'the barely concealed sexual excitement of its accelerating motion can hardly have disappointed the young engineers and their dancing partners'.[23] The journey gets off to a gentle start in the Introduction with tentative huffing and puffing in the lower registers, very gradually getting louder and gathering speed. The Introduction ends with two hoots on the train whistle, provided by the woodwind, helping to set the scene. Although *Accellerationen* starts off quietly, in contrast with the explosive start of the *Eisen-Bahn-Waltzer*, the younger Strauss adopts some of the same compositional techniques as his father such as tremolo strings and rising chromatic scale passages. The excitement builds up again in Waltz No. 1 which starts off quietly with trembling strings building up with furiously driving crescendo as the engine accelerates. Four more waltzes follow, each of them with powerful build-ups and crescendos, not least in the Coda which ends with a loud rising bass line, reinforced by snare drums and kettle drums towards the triumphant loud chords of the final cadence.

Accellerationen was not the only train-inspired piece by Johann Strauss II, he composed at least two more: *Spiralen* (Spirals, 1858); and *Vergnügungszug* (Pleasure Train Polka, 1864). For several seasons he took his orchestra to play in residence at the Vauxhall Pavilion[24], an elegant music pavilion and entertainment venue in Pavlovsk, the end station of the first 19-mile stretch of the Russian Tsarskoe Selo Railway. There they played ten successful (and well paid) summer seasons (1856-65) with further short visits later on. Strauss and his orchestra were one of the attractions, giving concerts every evening between May and September, encouraging well-to-do St. Petersburgers to use the new means of transport to travel there, thus helping to make the railway more profitable. Apparently these popular concerts were brought to an abrupt halt when the guard's whistle sounded for the last train to St Petersburg.[25]

The polkas

The polka originated in Bohemia in the 1830s and later became very popular as a ballroom dance in Vienna, largely thanks to the Strauss family. A lively dance for couples, it is performed with light hopping steps. Musically it has two beats to a bar and rhythms made up largely of quavers and semi-quavers in two-bar phrases. In Viennese ballrooms two distinct forms evolved, the graceful *Polka française* and the quicker *Schnell-Polka*. Eduard, who was often referred to as '*der schöne Edi*' (handsome Edi) had a successful musical career winning international acclaim as a conductor. In terms of composition, he was particularly famed for his polkas, carving out a niche writing in the hectic schnell polka style. His piece *Mit Dampf* (With Steam, 1871) is a typical example of the form. It is in ternary form (three sections) - Main polka theme, Trio and Finale. The first section of *Mit Dampf* has a fast and loud four-bar introduction (accompanied by a train whistle on some recordings) which heralds the main theme. The Trio then introduces a new theme using dotted notes and leaps. This is followed by the last section, the Finale. The Finale recaps the first section complete with its introductory train whistle heard on some recordings, and the rhythmic metallic sound as the train passes over the sleepers, cymbal crashes and plenty of percussion to add to the excitement. The polka dance ends with a final hoot from the train whistle.

Danger on the trains

Trains promised speed but also danger. These aspects of train travel are emphasised in the titles of two of Eduard Strauss's fast polkas: the raucous *Ohne Bremse* (Without Brakes) and *Ohne Aufenthalt* (Non Stop) which opens with a clanging bell in many performances and ends with a piercing (piccolo) guard's whistle. Instructions for first-time riders on the Nordbahn included prominent warnings against any attempt to leap from the car, which it suggests, occasionally happened when the steam engine's whistle blew.[26]

Train accidents across Europe and the USA were a common occurrence in the nineteenth century and newspapers were quick to report them.

> On the morning of 7 July, 1839, three trains operated by the Kaiser Ferdinands-Nordbahn, arrived at Brno/ Brünn…As one local newspaper correspondent wrote… In little time the first of three trains from Vienna pulled into the station. "With speed like the wind," it had covered the roughly 130 kilometers from the imperial capital to Brno in just four and a half hours. …That afternoon the three trains departed for Vienna, and within hours horrible news had reached Brno: the third train had slammed into the second train at a small station north of Vienna…a number of people had been injured, several of them seriously. Later investigations showed that the engineer of the third train had exceeded the company's speed limit, resulting in brake failure.[27]

The Nordbahn suffered several tragedies in the year that followed including a fire that broke out in the Vienna train station seriously damaging a number of cars.[28]

Pleasure trips

Danger or no danger, hundreds of Austrians were happy to take the many train trips on offer out of Vienna. The construction of the railways made trips into the countryside and fashionable spa towns not only much cheaper, and therefore no longer the sole preserve of the rich, but also possible to cover in one day. The first travel agent was opened in Vienna in 1866 and was run by Gustav Schrökl and his wife Therese. They organised a variety of train excursions: tours; trips to the theatre; trips to exhibitions; and pleasure trips. The *Morgen-Post* newspaper, reported that on one such pleasure trip from Innsbruck to Vienna in 1874, the passengers were cramped together, the food ran out and they had to endure a 36-hour return trip to Vienna referring to Schrökl's trips as 'The pleasure-seekers terror.'[29]

Several of the Strauss dance works celebrated pleasure trips including Johann Strauss II's *Vergnügungszug* (Pleasure Train Polka); and Eduard Strauss's *Tour und Retour* (Round Trip) and *Lustfahrten* (Pleasure Journeys). Johan Strauss II composed *Vergnügungszug* in 1864 for the Association of Industrial Societies Ball held in the ballroom of the Imperial Palace in Vienna. It uses the instruments of the orchestra to conjure up the sounds of the train; the piccolo represents the station master's whistle and syncopated horns signal warnings of the train's approach as it flies through the countryside. Johann Strauss II took the piece to Pavlovsk for his 1864 concert season. It was played at the opening concert and proved to be so popular that it was played many more times during his residency there.

Several other Strauss railway pieces were written for balls. In 1860 Josef Strauss composed the polka *Gruß an München* (Greetings to Munich), a *Polka française* for a party to celebrate the inauguration of the Vienna-Salzburg railway line, the Empress Elisabeth Railway, and much of Eduard's railway music was written for balls in Vienna commissioned by railway officials.

Bahn frei! Line Clear!

One such work was *Bahn Frei!* (Line Clear!, 1869). The piece, a quick polka, was written for the ball of the railway employees of the Nordbahn. Some joked that it was played at a much quicker tempo than the Nordbahn trains could actually travel at.[30] *Bahn Frei!* opens in imitation of a train with stuttering woodwind playing short quiet notes, getting louder and faster building up to piercing shrill high notes from the piccolo and flute - the guard's whistle announces the line is clear. The engine starts with a mechanical chugging rhythm on pizzicato strings, the polka melody begins and the full orchestra joins in as the journey gets underway. It is not difficult to see how the laboured departure of a steam engine, building up to a quicker development and finally a coda as the passengers reach the end of their journey is a handy template for composers to follow.

Johnson writes that some of the Strauss railway pieces were written

> in a form that became standard for all future 'train pieces' - beginning with the gathering of momentum from a standing start and ending by a corresponding wind down. In between the mechanical rhythm of the engine becomes the background accompaniment to melodic writing for the strings, suggesting the romanticized glamour of gliding through the landscape, elegantly motionless in one's own body even while the vehicle speeds along.'[31]

There is much that is true in this statement, but the Strauss railway pieces do not usually end with a 'wind down'. Bearing in mind that they were written in part to be danced to and that they acted as showpieces for the Strauss Orchestra, it is not surprising that rather than petering out, most end with a huge orchestral splash. The final bars of *Bahn frei!* are typical - a crescendo building up to fortissimo final chords with trills, crashing cymbals, and rolls on the drums.

End note

Some readers may wonder why there has been no mention of Johann Strauss II's *Reise Abenteuer* (Travel Adventures), a piece which is often included in recordings and lists of railway pieces. It would appear that these were not railway adventures, but adventures at sea. The cover of the first piano edition features a vignette of a paddle-steamer being tossed in a violent storm[32] and is a reflection of Strauss's own experience. In 1857 he had travelled by sea to Russia enduring a terrible storm. The impressive Coda is a dramatic musical depiction of storm-lashed vessels pitching and rolling, while the wind rages and the lightning flashes.

Endnotes

1. Hobsbawm, Eric. *(Industry and Empire: From 1750 to the Present Day.* (London: Penguin, 1969): 110-111.
2. Jeffrey Richards and John M MacKenzie. *The Railway Station: A Social History.* (Oxford: Oxford University Press, 1986).
3. České Budějovice in the present day Czech Republic.
4. Chad Bryant. 'Into an Uncertain Future: Railroads and Vormärz Liberalism in Brno, Vienna, and Prague', *Austrian History Yearbook*, 40 (2009): 187.
5. Christian Wolmar. *The Golden Age of European Railways.* (Barnsley: Pen & Sword Transport, 2013): 138-148.
6. Robert Barry. *The Music of the Future.* (London: Repeater Books, 2016): 89.
7. Peter Kemp. *The Strauss Family. Portrait of a Musical Dynasty.* (Tunbridge Wells: The Baton Press, 1985):16.
8. *Journal des Débats*, 10 November, 1837.
9. Orlando Figes. *The Europeans: Three Lives and the Making of a Cosmopolitan Culture.* (London: Penguin, 2020): 272.
10. Kemp, *The Strauss Family*, 155-157.
11. Kemp, *The Strauss Family*, 157.
12. Figes, *The Europeans*, 121.
13. Barry, *The Music of the Future*, 89.
14. Gustave Flaubert, *Madame Bovary*. (Minneapolis, Minnesota: Lerner Publishing Group, 2016): 58
15. Andrew Lamb. 'Waltz' in *Grove Music Online*.
16. The Russian diplomat Prince Razumovsky had lived at The Golden Pear for a few days, as had the playwright and novelist Honoré de Balzac. Ludwig van Beethoven was a regular customer between 1823-4 when he lived in the same area.
17. Julian Johnson. *Out of Time: Music and the Making of Modernity.* (Oxford: Oxford University Press, 2015): 102.
18. Kemp, *The Strauss Family*, 80.
19. Johnson, *Out of Time*, 102.
20. Johnson, *Out of Time*, 101.
21. Sigmund Freud. *The Complete Psychological Works Of Sigmund Freud, Vol 7: "A Case of Hysteria", "Three Essays on Sexuality" and Other Works.* (London: Hogarth Press, 1953): 202.
22. Johnson, *Out of Time*, 101.
23. Johnson, *Out of Time*, 102.
24. Named after the Vauxhall Pleasure Gardens in London.
25. Orlando Figes, *The Europeans*, 274.
26. Bryant. "Into an Uncertain Future": 195.
27. Bryant, "Into an Uncertain Future", 183.
28. . Bryant, "Into an Uncertain Future", 195.
29. Martin Scheutz, "Die Geschichte der Reisebüros" in *Fernweh und Stadt. Tourismus als städtisches Phänomen* (Innsbruck: Studien Verlag, 2018): 151-152.
30. https://wjso.or.at/de-at/Home/Events/EventDetail?ConcertID=813&WerkID=468
31. Johnson, *Out of Time*, 102.
32. Published by Haslinger in 1860.

6 The coming of the railways to Scandinavia, Hans Christian Lumbye and others

NORWAY AND SWEDEN both came late to railway building. In 1845 the Swedish count Adolf Eugene von Rosen had received permission to build railways in Sweden but his money ran out and the Swedish parliament decided that the trunk lines should be built and operated by the state. In 1854, the naval engineer Nils Ericson was employed to oversee the project. The first Norwegian line opened in 1854 (Christiana to Eidsvoll) and the first steam powered Swedish lines opened in 1856 (Nora to Ervalla) and 1857 (Arboga to Köping). The first two main lines were completed between 1860 and 1864 - the Southern, stretching from Stockholm to Malmö in the south, and the Western, which went to Gothenburg in the west.

When the Helsinki and Hämeenlinna line opened in 1862, Finland was then the Grand Duchy of Finland, a territory of Imperial Russia; consequently the railways were built to the Russian broad track gauge rather than the standard European gauge. By 1900 most of the main lines had been built, including the line to St. Petersburg.

Technically it could be said that the Danish railway began in 1844 with the opening of the Kile-Altona line in Holstein by King Christian VIII on 18 September 1844. However, as a result of the Second Schleswig War, Holstein was ceded to the German confederation in 1864 and it became a German railway although it had been built under the Danish monarchy. It could be argued that the true history began three years later with the opening of the line between the capital Copenhagen and the cathedral city of Roskilde. The Copenhagen-Roskilde line was inaugurated as the first in Denmark, on 24 June 1847. The construction was led by the English engineer William Radford and the Manchester-based Sharp Brothers and Company built the first batch of locomotives, the five ODIN locomotives. The original intention had been for the line to go from Copenhagen to Korsør on the west coast of Zealand and, when the funding had been secured, the line was extended to Korsør in 1856.[1]

One of the first Scandinavian railway pieces *Malmö Järnbanesång* (Malmö Railway song) was composed for the inauguration of the Lund-Malmö railway in 1856 by the Swedish composer Otto Lindblad (1809 - 1864). Fifteen years later it was a cause for celebration when Sweden and Norway were finally connected by rail and this was marked by Traugott Grahl in his piece *Sveas helsning till Nore, Walzer* (Greetings from Sweden to Norway, Waltz). The piece opens with a slow lyrical passage with woodwind and brass soloists in dialogue with the strings leading to the main waltz section which alternates between cheerful melodies and more melancholic passages.

Along with the waltz and the polka, the galop was one of the most popular ballroom dances in the nineteenth century. It is therefore no surprise that three of the pieces composed in celebration of the opening of Scandinavian railway lines were in this form. Jean Meyer, a violinist at the Stockholm Royal Opera House, wrote the *Jernvägs-Galopp* (Railway Galop) for the inauguration of the Stockholm to Gothenburg line. *Jernban-Galopp* (Railway Galop) by the little-known composer Franz Hoyer was composed for the inauguration of Finland's first railway and Hans Christian Lumbye's *Copenhagen Steam Railway Galop* was composed to celebrate the opening of the line from Copenhagen to Roskilde.

The galop is a fast, lively dance with two beats in a bar, its name comes from the galloping movement of horses. It was a simple but energetic dance to perform: dance partners galloped down the ballroom with springing steps. Because it was such a physically-demanding dance, galop pieces were usually quite short. Galops usually follow the same musical form which includes an introduction, trio and a finale (or coda) as well as the galop section itself. Each of the composers takes the opportunity to build up the excitement in the introduction as the train gathers speed and sets off. Meyer's *Railway Galop* opens with a rising figure on the strings, getting gradually faster before we hear the galop theme proper. He creates a convincing train imitation produced by sandpaper blocks. They are played with short strokes on the beat at first, getting faster as the train accelerates until a circular motion produces the sound of the train going at full speed – a process which is used in reverse as the train slows down at the end of the journey. Hoyer's opening music pictures the train starting up, gradually getting louder and faster as the train gathers speed, leading towards a light-hearted main galop theme. Both pieces make much use of flutes to suggest the sound of the train whistle, but what is unusual in Hoyer's train galop is an isolated dissonant passage where the flutes, glockenspiel and trilling triangles give way to strident brass and rumbling kettle drums in complete contrast to the carefree galop tune, perhaps emphasising the then inherent dangers of train travel. Unlike the pieces by Meyer and Hoyer which are largely forgotten, Lumbye's piece is still part of the orchestral repertoire and is explored in more detail.

Hans Christian Lumbye

The Danish composer Hans Christian Lumbye (1810 – 1874) had much in common with the Strauss family, in fact he is sometimes referred to as the Strauss of the North. He was a prolific composer of waltzes, polkas, galops and marches and made a career conducting his own orchestra in concerts of light music, mainly at the Tivoli Gardens but also, for one season at Pavlosk where members of the Strauss family had been such a success. Just as the Strauss family played a large part in creating the demand for light and accessible entertainment music, Lumbye was a very important figure in the creation of popular musical culture in Northern Europe.

He started his career playing trumpet in military bands and served in the Horse Guards in Copenhagen. Then in 1839 he heard an Austrian band playing pieces by Lanner and Strauss. This was the first time such pieces had been heard in Scandinavia and Lumbye was so impressed that in 1840 he formed his own light music orchestra and performed his first 'Concert à la Strauss' at the fashionable Raus Hotel in Copenhagen. He soon became a popular conductor in theatres around Copenhagen gaining a reputation as Scandinavia's undisputed leading dance composer. In 1843 the Tivoli Garden opened in Copenhagen. The now famous amusement park and pleasure gardens were originally named the Tivoli and Vauxhall, a reference to the Tivoli gardens in Paris and Rome and the London Vauxhall Pleasure Gardens. Lumbye was engaged as music director, in-house composer and leader of the concert hall's orchestra, a position he held for the next 30 years. There he composed about 700 dances, mainly polkas, waltzes and galops with several of them reflecting the connection with Tivoli such as *Tivoli Shooting Gallery Galop, Tivoli Bazaar*

Galop and *Tivoli Steam Merry-go-round Galop*. He also composed more than 25 ballet divertissements working in collaboration with the choreographer August Bournonville. Lumbye soon became a big draw to the Tivoli crowds, but also undertook a long series of tours visiting Hamburg, Berlin, Vienna, Paris, St Petersburg and Stockholm securing him an international reputation.

He remained a popular figure in the musical life of Copenhagen until his death in 1874. Two of his sons followed in their father's footsteps. Both Carl Lumbye (1841 – 1911) and Georg Lumbye (1843 – 1922) were successful conductors and composers of light music and both conducted the Tivoli wind band.

Kjobenhavns Jernbane-Damp-Galop (Copenhagen Steam Railway Galop) - Hans Christian Lumbye

Lumbye's *Copenhagen Steam Railway Galop*, a musical depiction of a railway journey, is one of his most famous and regularly performed compositions. The train sets off slowly out of the station, gradually accelerates, takes us on the course of the journey, and then slowly grinds to a halt. The piece is scored for a large orchestra with strings, woodwind and large brass and percussion sections. The orchestration includes several special train sound effects and, as the piece approaches the end, in performance some orchestral conductors announce that the train has reached its destination. Although performers of Strauss railway pieces often add their own train sounds, Lumbye's original score itself abounds with special effects. Towards the end of the Introduction there is part in the percussion section marked as *glocken geleute* (train bell) followed by an *Eisenbahn pfeife* (railway whistle). As the train starts to pick up speed in the next section the instruments chosen to imitate the train's rhythmic chugging are marked as 'sandpaper block plus springs plus bass drum'. Although sandpaper blocks and bass drum are not unusual choices to reproduce train sounds, the spring however is more unusual. The reco-reco is used to play this part today. It is a scraper of African origin, nowadays made of a metallic cylinder with springs attached and played with a metal stick. In Lumbye's time the spring specified would have been simply a real spring mounted with a resonator. In the final section we hear a conductor's whistle as marked in the score. As well as including these special effects, conventional orchestral instruments are also used to portray the sounds of a train: wide leaps and 'squeaks' marked in the flute, piccolo and oboe parts; long sustained notes in the horns; and rhythmic percussion parts.

The piece opens with a fairly slow Introduction, three in a bar, with a slow romantic melody on the cellos and oboe. Towards the end of the Introduction, twelve strikes of the train bell cut across the calm cello melody followed by five hoots on the train whistle and the music moves into a faster Allegretto section with two beats in a bar.

The sandpaper blocks, springs and bass drum imitate the acceleration of the train. As the train starts off they play on the first beat of the bar for four bars, then on both beats for two bars, then the two strokes increase to three, finally arriving at a regular four quavers in each bar, thus accurately encapsulating the acceleration of the train. At the same time the strings and woodwind play rapid semiquavers a semitone apart:

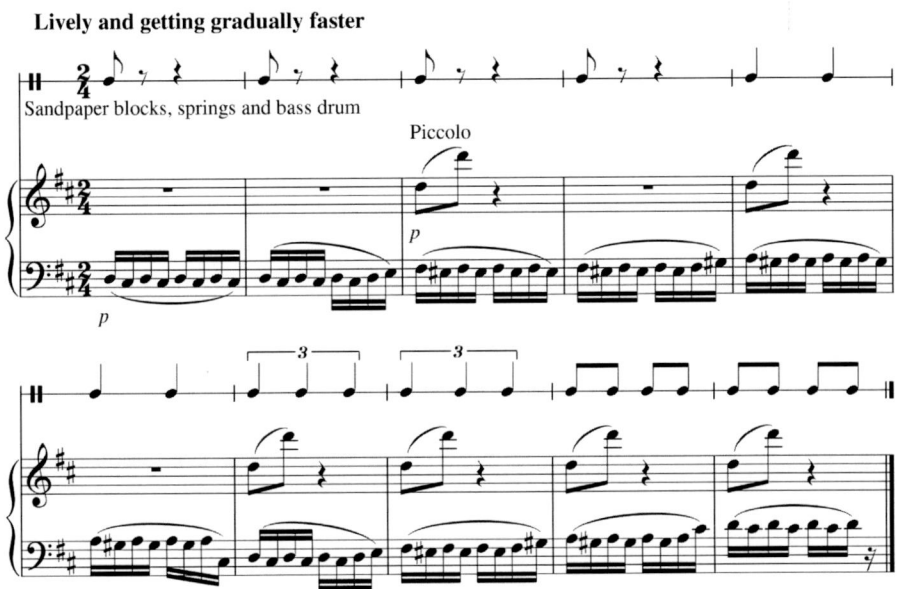

The music accelerates further to the fast main galop section based on this melody:

It then changes key for the slower moving Trio section:

Some listeners may be struck by the similarity the melody bears to one of the main themes in Julius Fučík's famous *Entry of the Gladiators*. The two pieces use a different time signature, but both melodies move upwards in a scale wise movement over four bars and then descend for two bars:

However, given the fact that the Railway Galop was composed in 1847 and the Czech composer Fučík was not born until 1872, any accusation of plagiarism cannot be levelled at Lumbye.

The final section, the Coda, opens with full orchestra and high swirling semiquavers in the woodwind and violins. As the train gradually slows down, the rapid semiquaver figure moves from the instruments' higher registers down to their lower registers, the tempo gets slower, the train whistle sounds and as the piece approaches the end, in some performances, the conductor announces that the train has reached its destination. The music gradually quietens down, with instruments dropping out in turn until only the lower strings are playing the semiquavers. The sandpaper block, springs and bass drum combination reappears for the final five bars and the music draws to a halt marked by a final loud chord played by the full orchestra.

Lumbye was a friend of the Hungarian composer and conductor Josef Gung'l and, although Gung'l was not Scandinavian, mention should be made here of his railway piece *Eisenbahn-Dampf-Galopp* (Steam Railway Galop) because of the similarity between this piece and Lumbye's *Copenhagen Steam Railway Galop*. Josef Gung'l (1809 – 1889) enjoyed a similarly popular career to Lumbye and the Strauss family, touring Europe and America with his light music orchestra as well as playing the summer seasons at Pavlosk between 1850 and 1855. He too was a prolific composer of dance music with many successful works under his belt.

Gung'l's *Eisenbahn-Dampf-Galopp* was instantly popular. A piano arrangement of the piece was published in Philadelphia, USA with the title 'Rail Road Steam Galop' and the added note 'as performed with unbounded applause'. A lithograph illustration of the railroad depot at Philadelphia is featured on the cover. The piano arrangement includes a stave for 'Locomotive Steam-Engine' with a repeated note on each beat of the bar, along with the instruction 'The noise which is produced by the steam in the Chimney of a Locomotive, may be very naturally imitated by clapping the Hand on a Stove Door.'

It is interesting to compare the ways that Gung'l and Lumbye depict the train as it sets off. Both use a percussive line to imitate the rhythmic chugging of the train. As we have seen, Lumbye takes care to build the train's acceleration into the rhythm of the percussion line (see page 93) whereas Gung'l sounds claps on the first beat of the bar for three bars, with a pause in the first bar, and then settles on the same rhythm for the remainder of the piece (see below). Another feature that both pieces have in common is the use of rapid semiquavers a semitone apart to emulate the movement of the train.

Lumbye would no doubt have heard of *Eisenbahn-Dampf-Galopp* and would certainly have been aware of its success. It had been played in Copenhagen by a visiting band from Steiermark in the spring of 1847 and it was not long after this that Lumbye composed his own railway galop. It could be argued that Lumbye was influenced by Gung'l's piece, although it should be noted that such rapid semiquaver movement is a common device found in many railway pieces to depict the movement of a train.

Endnotes

1 Christian Wolmar. *The Golden Age of European Railways*. (Barnsley: Pen & Sword Transport, 2013): 237-241

7 The coming of the railways to France, Charles-Valentin Alkan and Hector Berlioz

FRANCE WAS RELATIVELY slow to adopt railways in comparison with Britain, Germany and Belgium. The first railway to open in France was from St. Etienne westwards to Andrézieux in 1827. A limited service from St. Etienne north eastwards to Lyon then opened in 1832. It was in 1837 when the first full railway passenger service from Paris to St Germain began. A combination of reasons accounted for this delay: it was feared that railways would spoil the picturesque countryside; they would damage the water business industry; and on top of this the government was at first slow to make decisions. Under the rule of Napoleon III, state planning meant that by the end of the 1850s many companies had been merged and trunk lines were interconnected to form new networks. Six regional companies were established with five of them radiating out of Paris - Réseaux du Nord, Réseaux de l'Est, Réseaux de l'Ouest, Réseaux de Paris – Orléans, and Réseaux de Paris – Lyon– Méditerranée (PLM). The sixth network, Réseaux de Midi, served the south. This meant that long-distance travel often necessitated connecting to another line in the main hub of Paris. Consequently Paris grew dramatically in terms of population, commercial activity, industry and tourism. Two of the early trunk lines, Paris – Rouen (1843) and Paris – Orleans, had drawn on English technology and expertise, leaving a permanent legacy in the choice of gauge and the left-hand running which, apart from the Paris Metro, is still the practice on much of the double-track in France. The earliest locomotives in France were also of British design.[1]

Two notable pieces of railway music were composed in Paris during the early period of French railway construction: *Le chemin de fer* (1844), a rhythmically exciting piano piece imitating the sounds of a train by Charles-Valentin Alkan; and *Le chant des chemins de fer* a grand cantata for tenor and six-part chorus composed by Hector Berlioz to mark the opening of the Paris to Lille and Brussels railway line in June 1846.

Charles-Valentin Alkan

Although today Charles Alkan (1813-1888) is a relatively little-known composer, in the early nineteenth century he had a burgeoning career as both a composer and concert pianist which flourished alongside those of Liszt and Chopin (coincidentally his next door neighbour in Paris). Sadly his career did not thrive in the same way as theirs and his work gradually fell into obscurity. After his death, his music was admired by Debussy and Ravel and championed by pianists such as Busoni but it failed to attract a general following. However, since the 1960s, his work has attracted more attention leading to a small but enthusiastic following of aficionados.

Apart from some vocal works, Alkan composed almost exclusively for the keyboard. His 75 works for piano include the gargantuan *Twelve studies in all the minor keys, Opus 39*, which takes more than two hours to perform and includes a Concerto and a Symphony, both for solo piano. In contrast he also wrote numerous appealing piano miniatures, each portraying a different mood. As well as writing for piano, Alkan wrote several works for the pedalier, a piano fitted with pedals to be played by the feet in the same way as the organ. Unfortunately the pedalier is now obsolete, meaning that an important body of Alkan's music can no longer be heard.

Alkan was an erudite scholar with a wide range of interests including the study of theology. He was particularly interested in the Bible and completed the extraordinary feat of translating it from Hebrew into French. He had a great sense of humour which is evident in one of his wittiest and better known pieces - *Funeral march on the death of a parrot* - a piece for mixed voices and wind instruments which parodies the composer Rossini who had a penchant for parrots. Alkan had a life which was, in some ways, unconventional and often solitary; today's telling of it is surrounded by stories reinforcing ideas of his eccentricity. Some of these are true, others are not. The Alkan scholar Jack Gibbons has aimed to dispel some of the myths.[2] One of the most repeated is the story of Alkan's death which is commonly recounted as owing to a bookcase falling on top of him as he reached for a copy of the Talmud from the top shelf.

The truth is sadder and more prosaic. It seems that although he was trapped by a piece of falling furniture, it was probably a kitchen cupboard.[3] He never married and he lived alone, consequently it was more than 24 hours before he was discovered and dragged free but he died in his apartment a few hours later.

Le chemin de fer, Op. 27 (The railway) - Charles-Valentin Alkan

The first line for passenger traffic in Paris was opened on 24 August 1837 by Queen Marie-Amélie, wife of King Louis-Philippe. It ran from Paris to Le Pecq, the St Germain district north-west of central Paris, and was extended to Rouen in 1843. The terminal station in Paris was at *Embarcadère des Batignolles* which was within walking distance of rue Saint Lazare where Alkan lived at the time so this could be where he first experienced the railway.[4]

Le chemin de fer is thought to be the earliest musical representation of a journey on a steam-train. It had its first performance on 20 October 1844. The pianist was Alkan himself. The piece could be described as a *perpetuum mobile* – a continuous stream of notes played at a rapid tempo. It opens with a hypnotic regular beat as the train clatters along the tracks, then a stream of semiquavers enters as it hurtles onwards:

There is only one chord change in the first 52 bars. After 82 bars the relentless semiquaver pace eventually calms down giving way to a bucolic passage with a pastoral melody as the passengers survey the scenery outside. The pace picks up again and we hear a rumbling in the bass as the train belches steam:

It clatters along and then we hear the sound of a strident whistle in the piano's high register:

Towards the end of the piece, the train gets slower:

…and slower:

After a final whistle the train enters the station, it slows down and eventually comes to a halt at the journey's end:

In February 1845, *La France Musicale* reviewed the recently published piece making a comparison between performing the piece and going on a train journey with all its pleasures and dangers.

> We can…upon opening the first score, imagine ourselves sitting down, not at the piano, but in some carriage ready for departure. Can you hear the heavy, regular drumming of the steam train yet? Or, if you prefer, the persistent repetition of the regular figuration in the bass imitating the sound so well? The signal has been given and everything starts moving. You find yourself hurled in an uninterrupted succession of runs of semiquavers in the right hand. … do guard against wrong notes! The speed of the movement may derail you and you will not fail to realise the extent to which an accident might risk your reputation… while the landscape unrolls before our eyes – or at least, the wonderland of sounds captivates our enraptured ears – we will savour the delicacies of a sensual and seductive reverie: this *cantabile*.. so ingeniously divided between the two hands, invites us so gracefully to let our gaze roam from left to right… Then, ending our musing, we plunge again into the fervent whirlwind that carries us … The … *rallentando* announce, eventually, the end of our route and our étude. And so we have introduced our readers to the concept of imitative music.[5]

Le chemin de fer does indeed bring a certain sense of danger with its tremendous speed and relentless semiquavers - a reflection of the risks posed by train travel in 1844. The review above warns that 'The speed of the movement may derail you'. In 1842 France had experienced the worst rail disaster in the world at the time when a train travelling from Versailles to Paris derailed due to a broken axle on the leading locomotive.[6] The wreckage caught fire, killing as many as 200 people. In terms of the speed of the piece, some have suggested that Alkan's metronome marking was a mistake, being far too fast for anyone to play, and that the composer actually meant it to be half that speed. However there are virtuoso performances recorded at the original breakneck speed as indicated in the musical score above. Others have pointed out that at the time of composition trains would not have travelled at any great speed but it must be remembered that steam trains of 1844 would have far exceeded the speed of any other form of transport. Hammond estimates that the train would have been travelling at about 38 - 55 km/h (24 to 34 mph).[7] Although this would be pretty slow-moving by today's standards, it was much faster than the alternative means of transport at the time, the horse drawn carriage, which would have travelled at about 16 – 24 (10 to 15 mph).

Le chemin de fer was not the only composition of Alkan's to represent mechanics and technology. One of his earliest compositions, *Les omnibus* (1829), evoked the horse-drawn carriages commonly seen on the streets of Paris. Towards the end of the piece it includes a parody of the postillion's horn. Alkan also used the piano to evoke natural phenomena (*Le vent*, Op. 15 No. 2, *Comme le vent*, Op. 39 No. 1, *Neige et lave*, Op. 67 No. 1 and *Gros temps*, Op. 74 No. 10) and the animal world (*Saltarelle*, Op. 23 and *Le festin d'Esope*).

Hector Berlioz

Hector Berlioz (1803-1869) is now considered to be the leading composer of the Romantic movement in France. His musical ideas were highly original: he was a pioneer in creating new orchestral sounds; a master of using literature to create a musical narrative; and well-known by students of music as developing the 'idée fixe' where a melody is used to represent a person or an idea throughout an entire musical work – a concept which he developed in his most famous orchestral work, *Symphonie Fantastique*. During his lifetime however he struggled to find acceptance for many of his new ideas and failed to find a permanent and well-paid position in France. As a musician, he toured Europe a good deal but also earned his living much of the time through writing. He wrote countless articles of musical criticism along with several books notably a comprehensive study of orchestration, the *Grand traité d'instrumentation et d'orchestration modernes* (1843) and *Mémoires* (1870).

Le chant des chemins de fer (The song of the railways) – Hector Berlioz

Chant des chemins de fer, a huge cantata for tenor, choir and orchestra, was commissioned by the city of Lille to mark the launch of the Paris to Lille and Brussels railway line. It was first performed on 14 June 1846 in Lille.[8] The moving force behind the commission was the Lille judge Pierre Dubois who was a friend of the writer and critic Jules Janin, the author of the libretto. Berlioz received the commission on his return from a successful European concert tour. Whilst touring he had composed the bulk of a large-scale work for voices and orchestra - *La damnation de Faust* – but, keen to work on such a rare French commission, he dropped everything. He wrote to his friend and fellow music critic August Ambros.

> I am very busy with Faust, but I have just been forced to interrupt my work to write several feuilletons and a cantata which I am due to conduct in Lille for the celebration of the opening of the Northern Railway.[9]

Claude Monet (1840 – 1926) Train in the Snow (the Locomotive), 1875, Musee Marmottan, Paris. PRISMA ARCHIVO / Alamy Stock Photo

The work was written at great speed over a couple of days. Berlioz later observed that 'I'm quite aware that if I'd had three full days to devote to this piece, my score would live forty centuries longer.' He made several further references to the cantata's composition and performance in letters to his friends, but the most extensive and entertaining description can be found in his book *Les grotesques de la musique* (Musical Madhouse), a light-hearted collection of witticisms, letters and anecdotes about his life as a composer in mid-nineteenth century Paris.[10] *Le chant des chemins de fer* appears in the section of the book which describes various concerts of Berlioz's music given in the provinces. The description takes the form of a letter opening with the words 'I don't expect you have any desire to know why I went to Lille. I'll tell you anyway…The Northern railway, so celebrated for the little accidents to which it was prone, had just been completed.'[11] Here Berlioz was referring to a recent derailment near Arras where there were several fatalities.[12] He continues - 'His Grace the Archbishop was to give it his solemn blessing, which promised to be an occasion for copious eating and drinking. …a cantata was needed, to be performed, not after dinner, but before the opening of the ball.' The ceremony was to open with an outdoor military band performance of the 'Apothéose', the finale from Berlioz's *Grande symphonie funèbre et triomphale* (1840).[13]

As soon as Berlioz reached Lille, Dubois put him in touch with the singers, orchestral musicians and military bands who would be performing the piece. Berlioz was very impressed with the 'excellent small orchestra' and the young singers, writing that 'these young singers had excellent voices, and mastered the difficulties of the cantata in a very short time'. He was much less impressed by the massed military bands describing their playing as 'nothing but ear-splitting cacophony'. Much to Berlioz's amusement and surprise, Dubois also had plans to include pyrotechnics and cannon fire in the 'Apothéose' to 'produce a deafening effect'. In the event the first 'rocket shot up with such force it seemed to be heading for the moon' and the concert began. Another rocket shot up at the appointed bar in the coda in readiness for the cannon fire which would end the piece. Sadly the linstocks, the firing equipment for the cannon, hadn't arrive in time so 'A profound silence fell after the last bar, a majestic, grand, immense silence, disturbed an instant later only by the applause of the multitude, apparently satisfied with the performance' and quite convinced 'that the two rockets it had heard and seen whooshing up with showers of sparks were simply an new orchestral effect'. Following the calamitous outdoor performance Berlioz hurried to the town hall to conduct what was to be an excellent performance of *Le chant des chemins de fer*.[14]

> Our singers and musicians had not a breath nor a semiquaver to reproach themselves about. The same could not be said for the audience. After the concert, while I was listening to the gracious words the Duke of Nemours and his brother, the Duke of Montpensier, were good enough to say to me, some autograph collector did me the honour of stealing my hat. I was pained by this…I found myself compelled to go out bareheaded, and it was raining.[15]

Berlioz wrote of the verses of Janin's libretto 'they were fashioned in a certain manner which I shan't attempt to characterise, but which attracts music like ripe fruit attracts birds, quite unlike the great volleys of hemistichs[16] which professional librettists fire off'.[17] The composer appeared to be pleased with Janin's libretto, but Alastair Bruce, in the notes of his English translation of *Les grotesques de la musique*, is unequivocal when he writes 'His verses for *Le chant des chemins de fer*…were exceedingly banal.'[18]

Le chant des chemins de fer

Refrain: It is the big day, the feast day,
Day of triumph and laurels.
For you workers,
The crown is ready.
Soldiers of peace,
It is your victory;
The glory of so many blessings is yours.

Refrain

The bells ring at dawn,
And the cannon responds on the ramparts.
Under the tricolour banner
The people run from all sides.

Refrain

How many erased mountains!
So many rivers crossed!
Human labor, fertile sweat!
What wonders and what labour!

Refrain

The old men, in front of this spectacle,
With a smile will descend to the tomb;
For their children this miracle
Makes the future bigger, more beautiful.

Refrain

The wonders of industry
We, the witnesses, must sing
Peace! The king! The worker! The homeland!
And trade and its benefits!

It is the big day, the feast day,
Day of triumph and laurels.

May the happier people in the countryside so beautiful

By friendship raise their solemn voices

To God hidden in the heavens!

The piece, which lasts roughly 10 minutes, is not one of the composer's greatest works, but it is entertaining and fit for purpose; a festive celebratory work, written for large forces and triumphant in mood. It opens with a rousing introduction in a bright major key, loud and fast with rapid woodwind scales and fanfare-like brass heralding the entry of the solo tenor declaiming 'It is the big day' rising up and then falling:

Next the full six-part choir enter with the words 'For you workers, 'The crown is ready' with busy repeated quavers. The triumphant first refrain concludes with crashing cymbals and huge orchestral chords. Janin was a keen Saint-Simonian, as was Berlioz for a brief period during the 1830s.[19] Saint-Simonianism was inspired by the ideas of Claude Henri de Rouvroy, Comte de Saint-Simon (1760–1825) one of the founding fathers of French Christian socialism. Some writers, notably Pierre-René Serna, believe that Saint-Simonianism is evident in the libretto given that it includes some of the philosophy's key themes such as 'the wonders of industry', 'peace' and 'workers'. The refrain, he argues, could be interpreted as celebrating the 'big day', the advent of a new society, reflecting the Saint-Simonian ideology with its vision of an industrialized society where the workers were recognised and fulfilled.[20]

The mood changes with the third verse at the words 'The old men, in front of this spectacle, With a smile will descend to the tomb; For their children this miracle, Makes the future bigger, more beautiful'. In this calm hymn-like prayer, the music slows down, the time signature changes and the bass voices of the choir are now accompanied only by slow-moving strings. The fourth verse could be described as a *paean* to 'Peace! The king! The worker! The homeland!'. This section has a grandiose choral treatment where the soloist alternates with the choir in imitation, accompanied by trumpet fanfares, cymbal crashes and big bold chords. The whole piece builds up towards the end, with full choir and orchestra, punctuated by brass chords and cymbal crashes.

It has to be said that *Le chant des chemins de fer* is rarely performed. Notable performances in the United Kingdom include one in 1975 at the Royal Albert

Hall celebrating the 150th anniversary of the Stockton and Darlington Railway and another in 1994 at the launch ceremony of the Eurostar service. There are also few recordings. One interesting recording was made exactly 120 years after its composition when the French railway company SNCF organised a performance by the railway workers' symphony orchestra as part of the International Railway Congress Association conference in Paris.[21]

Endnotes

1. Christian Wolmar. *The Golden Age of European Railways*. (Barnsley: Pen & Sword Transport, 2013): 65-86
2. http://www.jackgibbons.com/alkanmyths.htm
3. Ibid.
4. Nick Hammond and Frederick Keygnaert. "Le Chemin de Fer". *The Alkan Society* Bulletin no. 96 (April 2018), 15.
5. *La France Musicale*, 9 Feb, 1845.
6. The accident happened on 8 May 1842 between Versailles and Paris following celebrations at the Palace of Versailles to mark King Louis Philippe's saint's day. 16 to 18 carriages carrying around 770 passengers, who were locked in their compartments, were being hauled by two steam locomotives. The leading locomotive broke an axle and derailed at Meudon. The second locomotive and the carriages behind piled into it and caught fire, seriously injuring hundreds and resulting in the deaths of between 52 and 200 people. Following this accident the custom of locking passengers in was abandoned.
7. *Alkan Society Bulletin*, no 96, April 2018
8. The piece exists in two versions: the original cantata for tenor and six-part chorus; and a later arrangement as No. 3 of his *Feuillets d'album*, Op.19.
9. http://www.hberlioz.com/France/Lille.htm
10. Hector Berlioz. *The Musical Madhouse*. Translated by Alastair Bruce. (Rochester: The University of Rochester Press, 2003).
11. This is the 'Third letter', an abridged form of a piece which originally appeared in the *Revue et Gazette Musicale* on 19 November, 1848.
12. According to the *Illustrated London News* (July 18, 1846) on July 8, 1846 a train from Paris derailed having recently passed through Arras station. The Northern Railway Company reported that 12 or 14 carriages derailed with five or six of them falling down a bank into deep water resulting in 12 fatalities. The ILN noted that other reports had put the number of deaths much higher.
13. Berlioz, *Musical Madhouse*, 177.
14. Berlioz, *Musical Madhouse*, 186-188.
15. Berlioz, *Musical Madhouse*, 188.
16. A hemistich is a poetic device, a half line of verse often separated rhythmically from the rest of the line by a caesura.
17. Berlioz, *Musical Madhouse*, 178.
18. Berlioz, *Musical Madhouse*, 209.
19. Berlioz, *Musical Madhouse*, 190.
20. Pierre-René Serna explores the elements of Saint-Simonian feeling in *Social Songs and Paths of Utopia*. http://www.hberlioz.com/Special/prserna.htm
21. The French railway company *Société Nationale des Chemins de Fer Français* (SNCF) invited the *Orchester Symphonique des Cheminots* and *La chorale de oratoire*, under the direction of Robert Blot, to perform the piece as part of the 1966 International Railway Congress Association conference in Paris.

8 The coming of the railways to Belgium, Gioachino Rossini

BELGIUM WAS THE first country in continental Europe to operate a steam railway, to create a national railway network and to possess a nationalised railway system. The first Belgian line connected its two main cities– Brussels and Antwerp. The line was completed in 1836 thus providing a route between the capital and the port without the need to use the inland waterways. The first lines and locomotives were imported from Britain and the first three locomotives to be used on the line were provided by George Stephenson's company. The completion of the first line was followed by a frenzy of railway building with lines extending to every main point of this small country creating the densest railway system in Europe.[1]

It may seem odd to feature the Italian composer Gioachino Rossini under the heading of the coming of the railways to Belgium. Furthermore, it is often stated that Rossini hated trains. However, many believe that his piano piece *Un petit train de plaisir comico-imitatif* (A little train of pleasure) which describes a catastrophic train journey was written at least partly in response to an unnerving experience he had when travelling from Antwerp to Brussels in 1836, its first year of operation. According to Macalpine and Hunter,

> he actually fainted during the journey and was so shaken that he was ill for many days after. He found it indeed so terrifying an experience that he swore never to set foot on a train again. And he never did, although this often meant lengthier and more arduous journeys by coach.[2]

We will never know what Rossini found so frightening, but we do know that Rossini was affected by nervous diseases throughout his life and that the train journey in 1836 would, at best, certainly have been uncomfortable; even for first-class ticket holders, the compartments were cramped and there was no heating or lighting and no access to toilet facilities or refreshments. Matters were worse for second and third-class passengers who travelled in open wagons. As Simon Bradley puts it, 'Facilities for even the most expensive travel in the earliest railway years went little further than padded seats…Illumination was provided by God's own daylight and the internal climate was governed by the external one'.[3] The ride was bumpy in the early days, the springs (if there were any) were weak and the track was uneven.

> Worse, the couplings between the coaches were not rigid…Every time the train set off or slowed down, passengers were thrown about. There was no brake mechanism to prevent the carriages bumping into each other and despite the padded seats in the superior carriages, complaints from the more well-to-do travellers about their uncomfortable journey were frequent and sustained.[4]

Nevertheless, although Rossini may have abhorred the idea of train journeys, he built his villa in Passy on a site close to the railway line and invested part of his fortune in railway shares.[5]

Gioachino Rossini

The career of the Italian composer Gioachino Rossini (1792-1868) falls into two halves. In the first part of his life he was hugely prolific, he composed rapidly and fluently, and by the age of 38 he had written nearly 40 operas, many of these highly successful. Often nicknamed 'Signor Crescendo' to reflect his talent for whipping up musical excitement, he had achieved international fame and recognition. Several of his operas - notably *The Italian Girl in Algiers*, *The Barber of Seville* and *Cinderella* - still form a significant part of today's standard repertoire. But Rossini's career as a stage composer spanned only nineteen years. In 1829 he composed *William Tell*, and this was to be his final opera, his career came to an abrupt end and he withdrew from society.

In 1855, 25 years later, he settled in Paris with his second wife, Olympe Pélissier. He bought a piece of land in the suburb of Passy where he built a villa as well as renting a place in the city quarters. He eventually re-emerged into society: his health improved; his joie de vivre returned; and remarkably he began to compose again. There were no more operas but there was a surge of composition, one large scale religious work the *Petite Messe solennelle* along with over 150 piano pieces, songs, and small ensembles that he referred to with the self-deprecating title *Péchés de vieillesse* (Sins of old age) (1857–68). Many of the smaller works were written to be performed at the Rossinis' '*Samedi soirs*' a series of elegant salons held at their two Parisian homes between 1858 and 1868, the year of the composer's death. The '*Samedi soirs*' soon became established as the most sought after artistic and intellectual gatherings and along with important public figures, the Rossinis were host to eminent composers, singers and instrumentalists including Wagner, Verdi, Liszt, Joachim and Rubinstein.

Un petit train de plaisir comico-imitatif (A little train of pleasure) - Gioachino Rossini[6]

The *Péchés de vieillesse* were written for Rossini's own amusement, to be performed in the privacy of his drawing room. He refused to allow them to be published and kept the manuscripts under lock and key in a mahogany chest in his bedroom.[7] They remained barely known until the Fondazione Rossini began editing them in the 1950s. Each of the piano pieces has a witty, often nonsensical title, such as *Prélude prétentieux* and *Valse torturée*. They are often humorous and sometimes mocking. Many of them are parodies of other composers or styles as revealed in their titles. In *Petit caprice* (style *Offenbach*) the tempo indication is 'Allegretto grotesco' and the music is reminiscent of a rather inebriated Offenbach cancan. Some of the pieces, including '*Un petit train de plaisir*' include an often satirical running commentary written on the score. The composer who is most well-known for peppering his scores with such whimsical instructions and witty remarks is Erik Satie (1866 - 1925) so there has been some speculation as to whether Satie was influenced in any way by *Péchés de vieillesse*. This seems unlikely given that Satie had probably not seen or heard any of this music when he was composing his own since the *Péchés de vieillesse* had not been published at this time.

What follows is a potted version of '*Un petit train de plaisir comico-imitatif*', a disastrous train journey translated into sound, what Richard Osborne describes as a 'not altogether amusing piano piece…in which genial mimicry of the train's progress is overtaken by a fatal derailment, the contemplation of the flight of the victims' souls and a sardonic coda in which the heirs of the more well-to-do victims cut a few celebratory capers.'[8]

The signal bell sounds and the passengers climb into the carriages:

Off goes the train, slowly gathering momentum:

A satanic whistle sounds:

Next we hear what Rossini describes (with a hint of sarcasm) as the 'Douce mélodie du Frein' (Sweet melody of the brakes). The music quietly runs down a chromatic scale, which could hardly be described as a melody, getting louder as it descends into the lower octaves then dying away and coming to a full stop and a pause as it arrives at the station. A slower more romantic passage follows with the comment: 'Les Lions Parisiens offrant la main aux Biches pour descendre du Wagon' (The Parisian Lions offering a hand to the Does to disembark from the carriage). This rather more obscure observation may have been a private joke for the salon gathering with the 'lions' referring to 'men' and the 'does' referring to 'women'. Just as in the wild lions prey upon does, perhaps the men are helping the women down from the carriages, but with wicked intent.

The journey continues but a terrible disaster is ahead. The convoy comes off the rails depicted by a chain of very loud and discordant diminished sevenths:

A first person is hurt and then a second. With a touch of irony, the first fatality rises up into Heaven with an ascending minor arpeggio whilst the second casualty goes down to Hell with a descending major arpeggio. Rossini has turned the standard major (happy)/minor (sad) convention on its head:

This passage leads into a more conventional funeral song with quiet slow-moving chords, many of them minor, bearing a noticeable similarity to both Chopin's funeral march and the slow movement of Beethoven's Symphony No. 7.[9] Perhaps this pastiche was another in-joke for the salon gathering that would no doubt have been familiar with both pieces.

Ex 24 Rossini (f)

The acute sorrow of the heirs, who will of course inherit the fortunes of those killed in the accident, is depicted with a note of cynicism by a carefree melody which is far from sorrowful. The piece continues in an exuberant vein, ending with exultant flourishes and loud C major chords. The irony is reinforced by the final observation *'Tout ceçi est plus que naîf: c'est vrai!'* (All of this is more than naïve: it is true).

Endnotes

1. Christian Wolmar. *Blood, Iron and Gold: How the Railways Transformed the World* (Great Britain: Atlantic Books, 2010): 21.
2. Ida Macalpine and Richard A Hunter. "Rossini: Piano Pieces for the Primal Scene" *American Imago*, 9, No. 3/4 (1952): 217.
3. Simon Bradley. *The Railways: Nation, Network and People* (London: Profile Books, 2015): 133.
4. Christian Wolmar. *A Short History of Trains*. (London: Dorling Kindersley Ltd., 2019): 147.
5. Macalpine and Hunter, "Rossini: Piano Pieces for the Primal Scene", 217.
6. Different versions of the title can be found for this piece. A comma is commonly used after the word 'plaisir', for example and sometimes the French 'comique' is used rather than the Italian 'comico'. It is not clear what is meant by *'comico-imitatif'*. It does not appear to be a genre as further examples of such a genre cannot be found. Sometimes it is translated as 'comic and imitative' no doubt referring to both the tongue-in-cheek comments and the music which, at points, imitates a train.
7. *The Times*, 27 December, 1968.
8. Richard Osborne. "Off the stage". In *The Cambridge Companion to Rossini*, edited by Emanuele Senici. (Cambridge: Cambridge University Press, 2007): 134.
9. The Marche Funèbre of Chopin's *Piano Sonata No. 2* (1839) and the Allegretto movement of Beethoven's *Symphony No. 7 in A* (1812).

9 The coming of the railways to Russia, Mikhail Glinka

THE RAILWAY NETWORK in Russia was slower to be established than in most parts of Europe. The first railway line was the result of a royal initiative; Tsar Nicholas I wished to connect two royal palaces, his main residence in St Petersburg and the summer residence, which had been favoured by Catherine the Great, in Tsarkoe Selo. The line was completed in 1837 and on October 30th the train was driven by Franz Anton von Gerstner, the Austrian engineer who had built the line 'with a huge audience of ministers and officials'.[1] Christian Wolmar writes of the opening of the St Petersburg to Tsarkoe Selo line that the first train 'carrying eight full coaches' took a 'mere 28 minutes to reach Tsarkoe Selo an impressive average speed of almost 50kph (30mph)'. The following year the line was extended by 16 miles beyond St Petersburg in order to connect it to Pavlosk, the fashionable entertainment resort. The line was a great success drawing the crowds through both their curiosity to see the new railway and their desire to experience the new attractions. The 400 mile line from St Petersburg to Moscow was completed in 1851 becoming one of the longest trunk routes in the world. Partly designed to act as defence barrier, the Russian lines were built to a broad track gauge rather than the narrow standard gauge found in most of Europe.[2]

'The travelling song' (1940) by the Russian composer Mikhail Glinka is often claimed to be the first railway song.

Mikhail Glinka

Mikhail Ivanovich Glinka (1804-1857), the first Russian composer of any stature, is best known for his two operas *A Life for the Tsar* (1836) and *Ruslan and Lyudmila* (1842). The son of an army captain, Glinka grew up in a wealthy family in Novospasskoye in the district of Smolensk. There he had his first contact with music listening to Russian folksongs and church choirs as well as performing himself in various local ensembles, mainly on piano and violin. His family were over-protective of him as a child and he was often cooped up inside rooms with too high a temperature. This may have led to his life of suffering from, and worrying about, his numerous ailments. In short he was a hypochondriac.

For much of his life he travelled around Europe with extensive stays in Italy, Germany, Poland, France and Spain. He counted many artists among his friends including the French composer Berlioz, the Italian opera composers Bellini and Donizetti, and the Russian writers Pushkin and Kukolnik. Although he was influenced by the music he came across, notably the music of Spain as well as Italian and French opera, his compositions were also imbued with Russian folksong and often used Russian themes as their subject matter. His first opera, *A Life for the Tsar*, came about as a result of his resolve to create an opera with a clear Russian sensibility, in the same way that the operas of his contemporaries Donizetti and Bellini were quintessentially Italian. It was premiered in the presence of the Imperial family in November 1836 and was an immediate success. Unfortunately that was not the case with his much-anticipated second opera *Ruslan and Lyudmila* which, with its more adventurous music, including Persian and Turkish influences, proved to be too exotic for the St Petersburg audience. Unsurprisingly Glinka was very disappointed and from then on focused less on composing and more on his European travels. Two years before he died he was persuaded by his sister to write the story of his life which is now entertainingly documented in his *Memoirs*.[3]

It is not clear when Glinka first travelled by train. He travelled widely, but in his *Memoirs* he makes scant reference to railway journeys although there is much attention paid to the types of carriages he travelled on by road, the many women he found attractive and the ailments he was currently suffering from. What he does write about journeys by rail would indicate that he was not greatly enamoured by that means of travel. In 1852 he travelled from Paris to Chalon-sur-Saône by railroad and he records that '…during the journey, I suffered for several hours from a nervous sinking or fainting sensation with feelings of terror'.[4] Similarly he writes that in the spring of 1853 he wished to return home from Paris but 'the railroad frightened me and I was trying to find some other way of travelling'.[5]

'The travelling song' No. 6 from A Farewell to St. Petersburg – Mikhail Glinka and Nestor Kukolnik

'The travelling song' is No. 6 of Glinka's cycle of twelve songs (or romances), *A Farewell to St Petersburg*. The cycle is written for voice and piano and has words by the Russian writer Nestor Kukolnik. According to David Brown, who wrote the standard text on Glinka, its title was 'prompted by Glinka's own wish to get away from the capital' at a time when he was experiencing various personal problems, not least the disintegration of his marriage. The complete volume of twelve songs was written quickly without much difficulty as Glinka writes in his *Memoirs*,

> The melody of the bolero *O beautiful maid of mine* came to mind …I asked Kukolnik to write some lines for this new melody and he agreed, at the same time offering me some other romances he had written…I had several spare melodies, so the work went along very well, I must say.[6]

According to Brown, in the case of 'The travelling song', the music was written before Kukolnik added the words.[7] *A Farewell to St Petersburg* provides a good illustration of what the *Grove Dictionary of Music* describes as Glinka's 'eclectic absorption of contemporary western techniques and idioms'.[8] Some of the songs have a Russian flavour, one or two have a Spanish influence whilst others have their roots in Italian music. Brown argues that 'The travelling song' owes something to Italian opera buffa (comic opera).[9] He has also been known to refer to it as an 'opera buffa puffer' and to describe it as a 'patter piece'. It is true to say that it displays several features of a patter song – a form often found in comic operas, particularly those of Gilbert and Sullivan. It has a fast tempo and a rapid succession of rhythmic patterns in which each syllable of the text corresponds to one note. The lyrics feature entertaining tongue twisting rhymes and the piano accompaniment is light and fairly simple in order to emphasise the words. It would also be true to describe it as a *perpetuum mobile* – a continuous stream of notes played at a rapid tempo.

The song is marked *presto* (very fast) and the tempo never wavers, travelling at top speed throughout, reflecting the speed generated by the locomotive engine. It is in a simple form (ABABA) where the refrain (A) alternates with the two verses (B). The opening refrain rattles away, two beats in a bar, in the bright key of D major as the train rushes through the countryside. The simple piano accompaniment has a straightforward rhythm and few chord changes with loud accented chords marking the end of each phrase.

The smoke is boiling, the steamer is smoking!
Diversity, revelry, excitement, anticipation, impatience!
Our Orthodox people rejoice
And faster, faster than will, the train rushes in an open field.[10]

In contrast with the refrain, the verse has a more lyrical character as the words tell of the anticipation of the journey's end. The minor key reflects the wistful mood of the words and the melody rises and falls against a background of colourful chords.

No, the secret thought flies faster,
And the heart beats counting the moment.
Insidious thoughts flicker along the road,
And you whisper involuntarily: How long, how long?

Finally – two myths to be dispelled. It is often written that 'The travelling song' was composed to celebrate the inauguration of the St Petersburg to Tsarkoe Selo line, but no direct link can be found to substantiate this. Furthermore it was written three years after the event. Nevertheless it is safe to assume that the new locomotives and the railway aroused much excitement at the time, in the same way as they had across Europe. The song is also often cited as being either the first railway song or the first piece of railway music, however, given that it was composed in 1840 this is not strictly true. In terms of songs, several British broadside ballads telling of railway building and railway openings were around by then (see Chapter 1) and one of the most famous Strauss railway pieces, *Eisenbahn-Lust-Walzer* (Railway pleasure waltzes), had been composed four years before. It might be more accurate to say that 'The travelling song' was the first art song to be written about the railway or indeed the first piece of Russian railway music.

Endnotes

1. Christian Wolmar. *The Golden Age of European Railways*. (Barnsley: Pen & Sword Transport, 2013): 217.
2. Christian Wolmar. *A Short History of Trains*. (London: Dorling Kindersley Ltd., 2019): 46-49.
3. Mikhail Ivanovich Glinka. *Memoirs*. Translated by Richard B Mudge. (Norman: University of Oklahoma Press, 1963).
4. Glinka, *Memoirs*, 231-2.
5. Glinka, *Memoirs*, 236.
6. Glinka, *Memoirs*, 150.
7. David Brown. *Mikhail Glinka. A Biographical and Critical Study*. (London: Oxford University Press, 1974): 159.
8. Stuart Campbell, 'Glinka, Mikhail Ivanovich' in *Grove Music Online*.
9. Brown, *Mikhail Glinka*, 159.
10. The song translation is by Peter Owens. As far as the different languages allow, it aligns the English words with their Russian equivalents in the original text.

Railway music in Europe and the USA in the twentieth century

This section of the book comprises five chapters covering music most of which belongs to the classical repertoire. The section which follows, 'Railway music in North America in the nineteenth and early twentieth century', covers music in a more popular vein.

The opening chapter examines three British railway pieces from the 1930s: *Night Mail* and *The Way to the Sea* by Benjamin Britten, and *Coronation Scot* by Vivian Ellis. The two composers worked in different fields; Ellis was a major figure in the field of light music and Britten went on to become a leading classical composer.

10 Three British railway pieces from the 1930s, *Night Mail*, *The Way to the Sea* and *Coronation Scot*

Night Mail and *The Way to the Sea*. Benjamin Britten and music for film

IN 1936 THE English composer Benjamin Britten (1913-1976) collaborated with the great poet W H Auden to work on two films about the British railway: *The Way to the Sea*, which celebrates the electrification of the Portsmouth to London railway line, and the much better known *Night Mail*. *Night Mail* tracks the journey of a Postal Special train from London to Glasgow documenting an everyday shift and emphasising the way in which the work of postal workers and railwaymen knit together in a remarkable act of collective organisation. As a marriage of film, music and poetry, the latter is still regarded by many as unsurpassed in its genre.

Britten was a young man at the beginning of his career when he joined the General Post Office (GPO) Film Unit where the poet W H Auden (1907-1973) was also a regular contributor. There Britten was expected to compose precisely timed scores to tight deadlines, an excellent grounding for the 22 year old composer. The GPO Film Unit was set up in 1933 by Stephen Tallents, partly to illustrate and promote the role of its quarter of a million employees. The short but distinctive films aimed to give a realistic picture of British society; they chronicled the working lives of British people, amongst them miners, fishermen, postal workers and railway men. The films were shown in cinemas, but more often in other venues such as factory canteens, schools and village halls and consequently reached a very wide audience.

The film unit was headed by John Grierson (1898-1972) whose vision was for a documentary approach to film making which, as the filmmaker Pat Jackson

(1999) wrote in his memoir, was 'To inform: to open eyes to new perspectives, new ways of thinking about social problems.'[1] Grierson was joined by like-minded others including the Brazilian-born film maker Alberto Cavalcanti, known for his experimentation with sound and music in film, and their combined output helped to enrich a tremendously innovative era in British film history. In 1939 the writer J B Priestley observed that 'Grierson and his young men, with their … taut social conscience, their rather Marxist sense of the contemporary scene always seemed to me at least a generation ahead of the dramatic film people.'[2] It is this underlying socialist ideology that has led to *Night Mail* being described as borrowing from 'the aesthetics of Soviet cinema to turn an explanation of the work of the travelling post office into a hymn to collective

Film poster for The Night Mail produced by GPO Film Unit, 1936
© Royal Mail Group Ltd 2021, courtesy of The Postal Museum

labour'.³ Grierson had written the English titles for *Turksib* (1929), a Soviet documentary directed by Viktor Turin, when it was released in the UK.⁴ The film documents the building of the Turkestan–Siberia Railway setting scenes of the open desert against the movement of machines and close-ups of pistons, wheels and machinery. As Scott Anthony writes, 'Turin's ability to evoke grand social developments with the Soviet's Spartan film-making resources proved inspirational to the similarly impoverished British documentary movement'.⁵

Night Mail was not the first GPO film about trains. In 1934 Humphrey Jennings' short film *Locomotives* about the development of the railway was largely made up of footage of miniature trains at London's Science Museum. It included extracts from Schubert's *Rosamunde* arranged by the composer John Foulds, but music was not a prominent feature of the film.

Night Mail was produced by John Grierson and directed by Harry Watt and Basil Wright. Britten and Auden's contribution was to supply the 'Title Music' and the 'End Sequence', roughly three minutes' worth of music to appear at the end of the 22 minute film. The remainder of the soundtrack is more focused on statistics - the millions of letters delivered each year, the number of points along the route at which mail is picked up and dropped off, and so on. Watt told Britten "Now I don't want any bloody highbrow stuff," urging him to make the music jazzy and emphasising the need for the music to be 'rhythmic to go with the beat of the train'.⁶ There is no discernible jazz influence in the finished product, but Britten famously composed music that, along with Auden's verse, captured the rhythms of a train. To familiarise himself with the different characteristics of trains sounds, Britten, a lapsed train spotter, enjoyed evenings at Harrow station accompanied by his film maker colleague Cavalcanti listening to the trains come and go.

A further restriction on Britten's brief was the necessity to match the sound exactly to what was happening on the film. At one point Britten wrote in his diary of the difficulties of writing "music" to 'minute instructions, when even the speed of the beat and number of bars is fixed'. An added difficulty was the need to fit Auden's words into the soundscape in strict rhythm with the music. Auden wrote the now familiar poetry at an old table in the noisy surroundings of the Film Unit's office in Soho Square, watching the rough cuts, timing the verse with a stopwatch and chopping and changing his lines accordingly. Britten chose the instrumentalists and conducted them in the recording – all for a fee of £13.10. Auden was taken on for £3.00 a week and was expected to supplement his poetic work by acting as Assistant Director. In 1936 Pat Jackson was a junior at the GPO Film Unit and was thrilled to be chosen as Harry Watt's assistant on *Night Mail*. He writes of filming 'some miles north of Hemel Hempstead' and how they 'trudged down the main "down fast" line to Crewe'.

> The L.M.S. railway company had provided us with a 'ganger' for our safety. He carried a red flag, had a whistle permanently between his teeth which he blew with monotonous regularity. The whistle would then drop suspended from a cord necklace and he would solemnly intone: 'Up fast. Stand clear', or 'Down fast. Stand clear'.[7]

Most of the film was shot as though it was a silent film; the sound was recorded later in various locations in a sound van.

Night Mail is scored for speaker and small ensemble – strings, harp, flute, oboe, trumpet in C and a percussion section (suspended cymbal, sandpaper, side drum, bass drum, and wind machine). The 'Title Music' lasts for 40 seconds. It opens with a roll on the side drum followed by a short, loud fanfare, seven notes on the trumpet with a typical fanfare rhythm. Britten had been instructed not to write 'any bloody highbrow stuff' but the choice of pitches for the fanfare is intriguingly difficult to analyse. Most fanfares are in a major key and clearly spell out the notes of a major chord, but this one has little sense of key and the chord it outlines is unconventional and difficult to classify. This could certainly be described as a nod to 'highbrow' modernity. However, Britten's gift for imitation and his ability to recreate the sounds and rhythms of the train ensured that the musical language was accessible to all. After a pause the music drops in volume and the flute, harp and strings have fluttering lines layered over each other. At the same time the percussion section imitates the sound of a train setting off with thuds on the bass drum which gradually become more frequent until the train sounds settle into a regular rhythm. Soon the sound dies away making away for the voiceover commentary.

The filming team travelled to Crewe where trolley loads of mail came in from all over the country. These were loaded onto the train and the dropped off through the night as the train travelled up to Scotland. Roughly half way through the film, shortly after the train leaves Crewe station, there is a 'Percussion Sequence' accompanying the commentator where the percussion is augmented by seven players producing special effects with 'found objects'. Britten had previously manipulated the sounds of found objects such as blocks of wood, chains and buckets of water, in *Coal Face* (1935), another Britten/Auden collaboration focusing on the lives of a Yorkshire mining community and their dangerous working conditions. The found objects in *Night Mail* are listed as

I   Steam (compressed air)
II   Sandpaper on slate
III   Rail
IV   Booms (clank)
V   Aluminium on drill/Motor Moy (A hand-cranked, chain-operated camera)
VI   Hammer on Conduit Boom/Siren
VII   Whistle/Coal falling down shaft

One such special effect reversed the sound of a hard beater on a light jazz cymbal to give the impression of a train whooshing through a tunnel. Listening to this sequence, it is hard to believe that the sounds emanate from a percussion ensemble rather than a train.

The shots of the 'Night Mail' climbing Beattock and descending to its destination in Glasgow were some of the last to be filmed. The footage was used to carry the opening lines of Auden's verse. The 'End Sequence' falls into four sections. The first section has a purely instrumental introduction where a wind machine and a muted side drum imitate the sounds of a train setting off in a similar manner to the Title Sequence. Once the stuttering train has fallen into a steady rhythm, the speaker enters with the famous words 'This is the night mail crossing the border'. Accompanied by side drum, the rhythm of the words is used to denote the propelling forward motion:

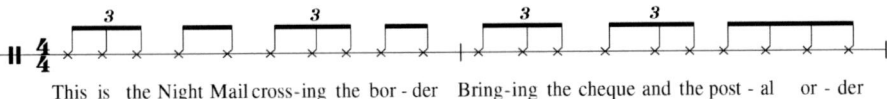

This is the Night Mail cross-ing the bor - der Bring-ing the cheque and the post - al or - der

Next the train climbs up Beattock Summit. This is reflected in rising figures of the strings and an upward leap on the bassoon. In musical terms this would be described as word painting where the music reflects the meaning of the words. The train gathers momentum flying past with a head of steam and the music builds up getting louder and more dense.

A climax is reached and the piece moves on to a new section, faster and with a change of key as the train descends into Scotland. It is on this leg of the journey that, set against an image of the driver wiping his face, we hear that the climb is over and the train descends into Glasgow. The speaker's pace is now more relaxed and the words less rhythmically tight. All the time the side drum beats out variants on the repeated train-mimicking rhythm:

The pace changes again in the third section, the train has built up speed and the music takes on a new urgency. The strings take on a chugging simple beat, but the words above have a rapid iteration listing the many and varied letters that the train is carrying; those of thanks, those from banks, and letters of joy. There are numerous words, multiple syllables with nowhere for the speaker, Stuart Legg, to breathe until he had a pause before the words 'Dreaming of terrifying monsters'. Britten wrote of this problem and how it was resolved: 'There is too much to be spoken in a single breath by the one voice (it is essential to keep to the same voice & to have no breaks) and so we have recorded them separately'.[8]

In the fourth and final section, with views of the city, we hear of the thousands dreaming, perhaps of monsters or maybe tea in one of the famous Glasgow tea rooms – Cranston's or Crawford's.[9] Next the words reflect on the way that the postman's knock quickens the heart with the fear of being forgotten. The music becomes quiet and reflective, only to build up again, the trumpet fanfares become prominent, but this time, rather than the obscure pitches in the opening, the fanfare spells out a triumphant chord of C major signalling the end of the night mail journey.

The premier of *Night Mail* took place in February 1936 at the Cambridge Arts Theatre. It received positive reviews across the national press and it is now acknowledged to be the most critically acclaimed film to be produced within the British documentary film movement. By focusing on the 'unusual and previously obscured skills needed to work at the Post Office' it 'proved extremely popular with both postal and railway workers'.[10]

The Way to the Sea

Although *The Way to the Sea* was written to celebrate the new electrification of the London to Portsmouth railway line, unlike *Night Mail*, trains and railway sounds do not feature prominently in the score. The documentary was made by Strand Films, and directed by John B Holmes. During December 1936, Britten and Auden worked together in what was to be their final collaboration, Auden, at one point, providing an extended verse commentary, and Britten composing the incidental music. It is scored for commentator, woodwind, brass, percussion, harp and piano. The musical style is rather different from that of *Night Mail*. There are no hints of modernism, no infectious rhythms in the text, there is little that is innovative, rather it relies more on musical pastiche. The piece has 11 movements and train references are made in only two of these. In the first of these movements the text excitedly boasts that the train time between London and Portsmouth has been reduced to 90 minutes along with 169 steam trains a week. Here mundane images are underscored by over-the-top pastiche ceremonial music, an unusual juxtaposition.

In the following movement, 'The line waits', Britten provides what can only be described as background music with some nods in the direction of train rhythm, but with Auden's commentary very much in the foreground. The newly electrified railway had opened up a commuter corridor between London and the south coast, and part of the intention of the promotional film had been to relate this to the lives of ordinary people. Auden's text is in parts bizarre and in parts supercilious. He writes of people who 'read adventure stories or understand algebra' or are 'brilliant at tennis' coupled with the need 'to keep respectable' and impress their neighbours. This rather strange choice of words juxtaposed with much use of pastiche has led some critics to believe that in *The Way to the Sea*, Britten and Auden had helped to create a satirical documentary under the guise of a promotional film.

Coronation Scot by Vivian Ellis

Vivian Ellis (1903–1996) was born in London, into a musical family. His grandmother, Julia Woolf, was a composer and his mother was a violinist. He studied composition and piano at the Royal Academy of Music, the latter under Myra Hess, with the intention of becoming a concert pianist, but he fell in love with the musical theatre while still a teenager. A crucial part of his training as a composer was his work as a reader and demonstrator for the London publisher Francis, Day & Hunter. This meant assessing songs and piano pieces submitted for publication - up to two hundred a week – as well as playing current publications in the shop to promote them. Next he began a long publishing association with Chappell & Co. and his first major success was *Mr. Cinders* (1929), several other operettas followed including *Bless the Bride*. His many hit songs included 'Spread a little happiness' and 'This is my lovely day'. During the 1960s and 1970s he faded from public prominence, but there was a resurgence of interest in his work when the singer Sting had a hit with 'Spread a little happiness' in 1982.

Ellis's music is melodic, sophisticated yet accessible, a style of popular music belonging firmly to the category of light music. Following in the steps of the operettas of Gilbert and Sullivan, and the waltzes and polkas of the Strauss family, light music pieces were composed to appeal to a wide audience. This made them ideal for broadcast throughout the heyday of radio. Many light music compositions are familiar as theme music, and it is for this reason that *Coronation Scot* is so well known. From 1947 until 1968, it was the signature tune for the BBC Light Programme detective series *Paul Temple* and it is safe to say that it would have been heard by millions.

The locomotive *Coronation Scot* has an interesting story. Its arrival made a big impact, but in its short two-year operating span, very few people had the opportunity to travel on it. The 1930s saw remarkable advances in railway technology with faster speeds and better comfort for travellers resulting in intense competition between companies. *Coronation Scot* was designed to complete the 401 mile journey from London to Glasgow in six hours and 30 minutes. It stood out from the crowd with its distinctive blue livery with silver stripes, and luxurious accommodation, and was an immediate success with the public. An opportunity arose to show it off in the USA and as a result a set of coaches, accompanied by 'Coronation' class locomotive 6229 *Duchess of Hamilton* (masquerading as 6220 *Coronation*) travelled 3,000 miles around the US before being exhibited at the 1939 New York's World Fair. However, within months the Second World War started, and the set had to be placed in storage in the USA. In the UK, the three original sets were placed in store where they remained until 1946 when they were repainted in conventional livery for general service.

Ellis wrote the *Coronation Scot* for Chappell's Recorded Music Library in 1938 and it was recorded by Sidney Torch and the Queen's Hall Light Orchestra. He was inspired by the rhythm of the train he travelled on between

Taunton and Paddington. This would have been the *Cornish Riviera Express*, but he settled instead for the name of the more prestigious train which was currently in the public eye.

The piece opens with the sound of a train whistle with loud chromatic chords, 'getting up steam' as is written in the score, the rhythm of the train is established by repeated accented chords, gradually speeding up until it reaches the familiar soaring melody on the strings. The journey continues replete with whistles, clanging bells and the occasional dissonant chord acting as a warning sign, until all quietens down and pizzicato strings bring the journey quietly to its end. Ellis was a master of both melody and orchestration, and it is the eminently hummable string tune that so effectively sums up an age of speed and luxury with the steam train in full flight.

Endnotes

1. Pat Jackson. *A Retake Please!: Night Mail to Western Approaches* (Liverpool: Liverpool University Press, 1999): 24.
2. J B Priestley. *Rain Upon Godshill* (London: William Heinemann Ltd, 1939).
3. *GPO Film Unit* (1933-1940). http://www.screenonline.org.uk/film/id/464254/index.html
4. The original music to *Turksib* was written by the Soviet composer Vissarion Shebalin (1902-1963). The film was released by the British Film Institute in 2011 as part of *The Soviet Influence: From Turksib to Night Mail* with a newly commissioned soundtrack by Guy Bartell.
5. Scott Anthony. *Night Mail* (London: British Film Institute, 2007).
6. Anthony. *Night Mail*, 41.
7. Jackson, *A Retake Please!*, 25.
8. Donald Mitchell. *Britten and Auden in the Thirties: The Year 1936* (Woodbridge, Suffolk: Boydell Press, 1981): 84.
9. The tea rooms Auden refers to are in Glasgow and Edinburgh respectively. Catherine Cranston (1849 – 1934) was an influential figure in the development of elegant high-quality tea rooms, some of these were designed by Charles Rennie Mackintosh.
10. Anthony, *Night Mail*, 84.

11 Railway music in Paris between the wars

THE FIVE COMPOSERS in this chapter - Arthur Honegger, Francis Poulenc, Darius Milhaud, Jacques Ibert and Heitor Villa-Lobos - all lived in Paris at some point in between the wars. At that time Paris was the place to be for musicians, writers and artists, so this is not surprising. What is more surprising is that three of them – Ibert, Milhaud and Honegger – were classmates, all studying composition at the Paris Conservatoire.[1] And on top of that another combination of the featured composers – this time Honegger, Milhaud and Poulenc – belonged to the group of composers who became known as *Les Six*.

The five railway pieces covered in this chapter are 'En chemin de fer' (1921), a movement from Poulenc's *Promenades* for solo piano, and 'Le Metro', a description of a ride on the Paris underground from Ibert's *Suite Symphonique* (1932) for orchestra. Milhaud's ballet music *Le Train Bleu* (1924) is also included although, as we shall see, the title is rather misleading given that the connection with this iconic train is limited. The chapter opens and closes with two of the most famous train-inspired orchestral pieces, *Pacific 231* (1923), by train-lover Arthur Honegger and 'The little train of the Caipira' by Heitor Villa-Lobos.

The inter-war years saw Paris as the intellectual capital of the world for artists and writers as well as musicians. It was not just the French who were drawn there. Villa-Lobos left his native Brazil for Europe in 1923 and eventually settled in Paris. The American composer Aaron Copland moved there in 1921 to study with the eminent composition teacher Nadia Boulanger and later wrote that 'Paris was filled with cosmopolitan artists from all over the world, many of whom had settled there as ex-patriots'. He goes on to list art movements such as Dada and surrealism, writers including James Joyce, T S Eliot, and Marcel Proust. He added that painters too were 'enormously active, with Picasso taking centre stage'.² Stravinsky could be added to this list of important figures, having taken up residence in Paris in 1920. The entry of the United States into the war in 1917 saw an influx of Americans into Paris, early jazz arrived with them and the city remained host to many jazz artists. It was a period where composers 'sought new music to voice the new age'.³ According to Milhaud, 'At this time everything was possible, we could try everything we wanted: it was a period of experiment, of liberty of expression in the widest sense of this word'.⁴

Les Six

The members of *Les Six* were the aforementioned Honegger, Milhaud and Poulenc along with Georges Auric, Louis Durey, and, the only female member of the group, Germaine Tailleferre. The term *Les Six* was really little more than a journalistic label. Most of them admired the music of Satie and the work of their mentor Jean Cocteau. They were against Wagner's influence and that of Debussy, but otherwise they had little in common; they were a group of disparate personalities composing in different styles.

Arthur Honegger

As a boy Arthur Honegger (1892-1955) loved to walk past the railway yards in his home town of Le Havre and he is oft quoted as saying "I have always loved locomotives passionately. For me they are living creatures, and I love them as others love women or horses." He was a student at the Paris Conservatoire from 1911 to 1918. In 1926 he married fellow student Andrée Vaurabourg on condition that they lived separately because he needed solitude for composing. He was a prolific composer of symphonies, operas, ballets and oratorios along with more utilitarian music for film and radio. When he died of a heart attack in 1955 he was given a state funeral by the French government. Although he collaborated with Ibert on the opera *L'Aiglon* his membership of *Les Six* can be misleading. In some ways his music had more in common with German romanticism; he made a significant contribution to the traditional forms of symphonic and choral repertory and counted Bach and Beethoven amongst his heroes.

Pacific 231

Honegger's symphonic movement was composed in 1923. Its original title was *Mouvement Symphonique No. 1*. The number 231 refers to a steam locomotive with two front axles, three main axles and one axle at the back. In his description of *Pacific 231* the American composer Aaron Copland writes

> Honegger took advantage of the fact that there is a certain analogy between the slow starting of a train, its gradual pickup of speed, its rush through space, and the slowing down to a full stop – and music. He manages very well to give the listener an impression of the hissing of steam and the chug chug of the mechanism and, at the same time, to write a piece solidly constructed of melodies and harmonies like any other piece.[5]

George Revill in his book *Railway* writes of the train as a 'symbolic force for expressing mechanical energy' outlining the way that the 'rhythms of mechanical motion' are gathered together into 'one single reverberating propulsive energy'.[6] This brings to mind the Futurist movement which sought to express the spirit of modernity in mechanical imagery, paying homage to and even imitating machines. Futurism was popular in the early twentieth century and Honegger had a certain affinity with it. *Pacific 231* could be said to have been created in this spirit. Its hard-edged sounds evoke raw power as the machine starts to move, gains speed, and then races with a sense of inexorable machine-driven motion. But was it Honegger's intention to imitate the sounds of a train journey? In Honegger's own word's

> What I have endeavoured to describe in *Pacific 231* is not an imitation of the sounds of the locomotive, but the translation into musical terms of the visual impression and the physical sensation of it. It shows the objective contemplation, the tranquil breathing of the engine in repose, the effort of starting [and] the progressive gathering of speed ... of a train of 300 tons hurling itself through the night at 120 miles an hour.[7][8]

He also described the piece in more musical terms

> In Pacific I was after an extremely abstract and wholly ideal notion: that of giving the impression of a mathematical acceleration of rhythm while the speed itself decreased.[9]

Both readings are valid. *Pacific 231* undoubtedly evokes the sounds of a train journey as is evident in its effective use as the soundtrack to Jean Mitry's 1949 film *Pacific 231*. As the American music critic Olin Downes wrote in his review of the first New York performance, 'This is as successful a balance of reaistic and genuinely musical ideas as we have encountered in the concert room...we have more than the imitation of a noise'.[10] The sounds of the train are produced, at least partly, by Honegger's mathematical approach. His metronome markings show that after the slow beginning, which is largely made up of sustained chords, the speed of the overall beat decreases from 160 beats per minute (bpm), down to 152 bpm, then 144 bpm, then 138 bpm until the final 13 bars where it slows right down to 126 bpm. At the same time the note values get smaller meaning that there are more notes played on each beat giving the impression that the music has speeded up.

In order to conjure up the unusual sonic landscape, Honegger has written for an orchestra including a rich palette of woodwind and brass instruments. A piccolo, cor Anglais, bass clarinet and contra bassoon are added to the usual woodwind section giving a wider range of sounds particularly in the bass register and the brass section includes three trumpets, three trombones, four horns and a tuba. There are also four heavily-employed percussionists. Furthermore Honegger includes a range of playing techniques and effects including flutter tonguing in the flutes, oboes, horns and trumpets.

The 'tranquil breathing of the engine' is portrayed by the slow chordal opening. This is built upon a foundation of trills in the double basses with the remaining strings playing with the bow near the bridge of the instrument producing a glassy, almost ghostly, tone evoking the sound of steam escaping. The 'effort of starting' with the great weight of the engine is suggested by the cumbersome block chords played by the low instruments of the orchestra followed by an insistent, accelerating rhythm with the stridency of the orchestra increasing as the music gathers momentum. The horns introduce the first main theme which is soon taken up by the trumpets in shorter (quicker) note values and then passed around the orchestra. Next a second more aggressive theme is heard on the bassoons and makes it way upwards through the orchestra. Off-beat, sometimes conflicting rhythms go back and forth from different parts of the orchestra in ever-new patterns and combinations, all punctuated by insistent drumming. As the train picks up speed, the note values become shorter and a new swirling tune is heard in the higher woodwind. At the height of the train's journey, there is a heady sense of gathering momentum, what Honegger describes as the 'progressive gathering of speed ... of a train of 300 tons hurling itself through the night'. A new broad theme is played loudly on the horns and eventually combines with other themes in a shattering chaotic climax. At length everything comes together in raucous loud chords with the brakes applied as the music slows shudders to a halt with its emphatic final chord and, some would say, the train reaches its destination.

Reception of *Pacific 231*

The first performance took place in May 1924 at the Opéra in Paris. It was met with great anticipation, the house was packed and it was an instant success. Furthermore it stirred up much excitement amongst concert goers across Europe and North America, everyone was keen to hear it. However, there was much debate about the artistic merits of a piece of music depicting a train.

The Illustrated London News review of the Royal Philharmonic Society performance under Eugene Goossens in 1925 argues that 'To give a musical imitation of a railway engine might be funny, but it would not be art as we know it today'.[11] *Truth* magazine reported that the performance 'stirred the usually apathetic Philharmonic audience to boos and hisses mingled with applause' but concluded that the 'musical impression of a locomotive' was an 'exceedingly effective one'.[12] Later that year the *Westminster Gazette* reported a full house at the Proms and that 'Honegger's sensational – or would be sensational – "Pacific 231" was another "big noise" both literally and figuratively' which had helped to draw the audience.[13] It went on to say that the performance was 'much more suggestive of an engine which wants oiling and won't go than of the sustained and thunderous power which the alluring title leads one to expect'. Similarly Charles O'Connell in his 1935 *Victor Book of the Symphony* described it as a 'rather cheap imitation of a locomotive and its strident dissonances'.[14]

A much-anticipated first American performance was given at the Carnegie Hall in 1924 by the New York Symphony Orchestra under the baton of Walter Damrosch. Olin Downes reported in the New York Times that

> The concert "attracted numbers of the curious who came to Honegger's piece

and promptly left when it was over, though there were such masterpieces as Beethoven's Fifth Symphony, and Vaughan William's Fantasie on a Theme of Thomas Tallis...also to be heard. They did not matter. The people wanted to hear the locomotive, and even blasé critics, troubled by many concerts, did leg work so that they might arrive on time to describe the much advertised tone-picture to an expectant public.[15]

Downes goes on to give this eloquent description of *Pacific 231*, the last words on the piece should go to him.

And so there is the suggestion of the monster in repose, just breathing; then the start, in heavy chunking chords low in the orchestra; then the gradual acceleration, and generation of a fragment of joyous song. This is as successful a balance of realistic and genuinely musical ideas as we have encountered in the concert room. The suggestion of it is unmistakable, and is accomplished with expert technic, with a well-developed harmonic idiom and a thorough knowledge and ingenious employment of the orchestra. But we have more than the imitation of a noise. We have music, youthful, energetic, full of laughter. The units of rhythmic energy cohere and develop in the most organic manner to the moment when great smashing chords bring the end. The composition need not be taken portentously or as the discovery of a new phase of art: it is rather a highly amusing "jeu d'esprit".[16]

'En chemin de fer' from Promenades by Francis Poulenc

Born into a wealthy Parisian family, Francis Poulenc (1899- 1963) stormed onto the scene of Parisian culture as an 18 year-old composer and pianist. One of his best-known works is the satirical ballet *Les Biches* (1924) commissioned by the ballet impresario Diaghilev for the Ballets Russes and an instant popular and critical success. Poulenc experienced a religious epiphany in the 1930s and a series of sacred pieces followed including his *Gloria* (1960) and *Stabat Mater* (1951). His compositions are quite different from those of Honegger. Much of Poulenc's music has a simplicity and directness, elegant and melodic, it is often witty and, at times, flippant.

'En chemin de fer' is the eighth movement of Poulenc's suite of ten short pieces for solo piano. Each of the pieces portrays a different means of travel, starting out on foot (*A pied*), then by car (*En auto*), next on horseback (*A cheval*) arriving at a lake and continuing by boat (*En bateau*). A plane is boarded (*En avian*) and on landing the journey continues by bus (*En autobus*) then horse and carriage (*En voiture*) followed by a train journey (*En chemin de fer*). A bicycle is taken at the station (*A bicyclette*) and then a trip home by stage coach (*En diligence*).

Poulenc was only 22 when he composed *Promenades* and he was still studying music. *Promenades* could be considered as an experiment in different styles. Each of the pieces employs a different compositional technique, resulting in heightened levels of dissonances, sometimes verging on atonality. Artur Rubinstein, to whom the *Promenades* are dedicated, first performed them at the Wigmore Hall London in 1923. *The Musical Times* reported in its review that 'The Poulenc Suite is far too clever, but has charming moments and a pleasant if over-conscious humour.'[17] 'En chemin de fer' could well be described as such 'a charming moment'.

At first sight the music for 'En chemin de fer' seems simple. The opening bars use simple chords, regular rhythms and there are no flats or sharps. However, the tempo (*vif*) is very fast, as the piece goes on the music passes through several keys and there are some awkward leaps in the left hand, making it rather more difficult to play than it first might appear. There are no attempts to imitate a train in this piece other than the sense of movement as the train speeds along on its journey.

Darius Milhaud

Darius Milhaud (1892-1974) was the most avant-garde member of *Les Six* and, with over 4000 compositions one of the most prolific composers of the twentieth century. Many of his compositions are influenced by jazz and Brazilian music. From 1917 to 1919, Milhaud worked in Rio de Janeiro as secretary the French ambassador to Brazil. Whilst there he formed a friendship with Villa-Lobos who introduced him to the music of local street musicians, Milhaud's most famous piece is his surrealist ballet *Le Boeuf sur le toit* which is named after a Brazilian popular song.

Le Train Bleu

The title of this 1924 ballet is misleading; although it refers to the luxury express train operated between Calais and the French Riviera, *Le Train Bleu* is never actually seen on stage. Its fashionable Parisian passengers have already disembarked at the Côte d'Azur. It has been included here for two reasons: it is often cited in lists of railway music and, although it is little performed, it is an intriguing piece created by leading artists of the time (Picasso, Coco Chanel and Cocteau amongst them). As Diaghilev explains in his programme notes

> The first point about *Le Train Bleu*, is that there is no blue train in it. This being the age of speed, it has already reached its destination and disembarked its passengers. They are to be seen on a beach which does not exist, in front of a casino which exists still less. Overhead passes an aeroplane which you do not see. And the plot represents nothing [...] the music is composed by Darius Milhaud, but it has nothing in common with the music we associate with Darius Milhaud.[18]

Jacques Ibert

Jacques Ibert (1890 – 1962) had an interesting and varied life. At the Paris Conservatoire he won the prestigious *Prix de Rome* despite his studies having been interrupted for four years by his service in World War I, first as a stretcher-bearer at the front, then as a naval officer stationed at Dunkirk. World War II was a difficult period for him; in 1940 the Vichy government banned his music and for a time he went into exile in Switzerland. In 1955 he was put in charge of both the *Opéra* and the *Opéra-Comique* and shortly afterwards he was appointed as the director of the *Académie des Beaux-Arts*. Through all this he had a successful and prolific composing career. His music embraces a wide variety of styles and moods as is evident in his two best-known compositions - the witty and frivolous *Divertissement* for small orchestra and the more serious romanticism of his large-scale orchestral piece *Escales*.

'Le Metro'

'Le Métro' is the first movement of Ibert's *Suite Symphonique*, sometimes known as the *'Paris Suite'*, (1932). The suite is written for a small orchestra with the unusual additions of piano, harmonium and celesta. It has a large percussion section including xylophone, two glockenspiels, wood block, and tam tam. In Ibert's words, taken from the top of the score, he wished to 'express musically the different faces of Paris'. This rather eclectic mix of scenarios, as Ibert explains' is 'from the score of Jules Romain's play "Donongo" which was showing at the Théâtre Pigalle in Paris around the same time.

I. Le Métro (The Metro)
II. Faubourgs (The suburbs)
III. La Mosquée de Paris (The Mosque of Paris)
IV. Restaurant au Bois de Boulogne (Restaurant in the Bois de Boulogne)
V. Le Paquebot "Île-de-France" (The steamship "Île-de-France")
VI. Parade foraine (Parade at the fair)

The Paris Métro was relatively new when the piece was written. The first line opened in 1900 during the Exposition Universelle, the World Fair. It expanded rapidly under the supervision of chief engineer Fulgence Bienvenüe and the main core was completed by the 1920s with extensions into the suburbs following in the 1930s. The Art Nouveau station entrances were designed by Hector Guimard. From its beginnings the Parisian underground lines carried a large number of passengers. Ibert describes 'Le Métro' at the top of the score as follows

> Eight o'clock in the morning. The crowd crushes toward the platform. The trumpet sounds the signal for departure. The train gets under way. The horde sits passively as the underground journey begins.

Nowadays the Parisian Metro is much quieter than it was in 1930 when many of the trains used were made completely of metal including steel wheels, but these old trains have now been replaced by modern models which have a quieter sound.[19] The soundscape of Ibert's train is one of roaring, clanking, rattling dissonance. It opens with a rumbling roll on the kettledrum under a huge cacophonous chord rising up to trills in the higher woodwind, they fade out, a trumpet announces that the train is about to set off followed by a piano and glockenspiel imitating the sound of train bells. There is a pause before the main fast section begins, opening with a brief trumpet fanfare followed by a strange train hooting sound effect from the clarinet. The clarinettist is instructed to 'blow into the mouthpiece' which involves first detaching it from the instrument producing a rather unearthly sound.

It could be argued that, like *Pacific 231*, 'Le Metro' too has elements of the Futurist aesthetic in the way that it expresses mechanical imagery. As well as using train sound effects it also evokes machine-driven motion through musical means. Once the train gets going its mechanical energy is captured through a repeated figure (ostinato) played by piano, snare drum and bass drum. Underneath this percussive ostinato the strings play another repeated figure which makes a veiled reference to George Gershwin's jazzy orchestral piece *American in Paris* which had recently received its Parisian premiere.

The train powers along getting ever noisier, we hear a warning sound (long loud notes on the trombone) as we approach the next station, the music gets louder until we come to an abrupt halt. We are left with low sustained string chords, a few clicks and creaks from the percussions, and some isolated squeals from the high woodwind in the final bars – the kind of sounds heard when a train comes to rest in an underground station. The whole journey is short, lasting only about two minutes - it is the underground after all.

Heitor Villa-Lobos

Villa-Lobos (1887-1959) was raised in Rio de Janeiro. His father was a keen amateur musician who encouraged his son's musical interests. Under his guidance, he played the clarinet and soon became an accomplished cello player, as a teenager he also enjoyed playing Brazilian popular music on guitar with Rio's street musicians. In 1903 he left home, earning a living mainly by playing the cello in theatres and hotels. His fascination with Brazilian popular and folk music led him to make various trips around Brazil and the Amazon where he learnt dozens of songs and tunes. These were brought into play in many of his compositions, helping to create a vibrant new sound.

As a composer, he was self-taught and prolific. It has been written that he dashed off his compositions in 'feverish haste (often to the accompaniment of radio, conversation, and other music in the house).[20] By the age of 30 he had produced over 100 works and was well-established as a composer of Brazilian art music. In 1923 he left for Europe, a trip which was subsidized by wealthy friends and a government grant. He settled in Paris moving in artistic circles with his friend Milhaud, who he had first met in Brazil, along with Ravel, Stravinsky and Prokofiev amongst others. Whilst living in Paris he put on some highly successful concerts of his own music. In 1930 he moved back to Rio where he presented a plan for music education to the State Secretariat for Education. Later that year his plan was backed by the new government under Gétulio Vargas and he instigated a nationwide programme of music instruction for schools using Brazilian popular music. He also organised choral singing on a mass scale. On one occasion in 1935 this involved some 30,000 voices and 1000 band musicians.[21]

One of the major instrumental landmarks in his output is the set of nine suites *Bachianas Brasileiras* for various combinations of voices and instruments. These are pieces which combine the sounds of Brazil with contemporary classical music techniques and, as the title suggests, elements of the style of the Baroque composer J S Bach. Although sharing a music language found in the works of some of his contemporaries, Villa-Lobos was unquestionably a nationalist composer and these eclectic works are permeated with the instruments, rhythms and melodies of the folk and popular music of Brazil.

'The little train of the Caipira' from *Bachianas Brasileiras No. 2*

'The little train of the Caipira' is the final movement of the four-movement orchestral suite *Bachianas brasileiras No. 2*. It is a short piece lasting around four minutes. Although the suite was completed in Rio de Janeiro in 1930, it is safe to assume that the process of composition began in Paris.[22] Each of the movements has twin titles, one of which alludes to the movement of a Bach suite (in this case the Toccata) and the other to some facet of Brazilian life. The word 'Caipira' refers to the inhabitants of rural areas in the southern interior of Brazil.

Bachianas brasileiras No. 2 is scored for a large chamber orchestra of ten wind and brass instruments, a string section, piano and celesta, and a percussion section (timpani, triangle, cymbals, tam-tam, snare drum and bass drum) supplemented by South American instruments. The music of Villa-Lobos is noted for its interesting textures and vivid orchestration, and the use of train sound effects and South American percussion in 'The little train of the Caipira' is an exemplar of both. Much of the colour in this movement comes from the composer's use of the ganza (metal tube filled with gravel - shaker), chocalhos (metal tube filled with beads- rattle), raganella (ratchet), reco-reco (notched stick - scraper) and tamburello (tambourine).[23] It is these instruments which, from the opening bars of the piece, add so much to the evocation of the rhythmic sounds of the steam locomotive.

The piece opens quietly as the train gets off to a wheezing start, the strings play pianissimo sustained chords whilst the percussion sets the train in motion with its repeated rhythmic patterns. The steam whistle (flute and clarinet) sounds, and the train gradually picks up speed. The scraping and shaking of the South American instruments coupled with the percussive block chords on the piano combine to make a very convincing steam train. As the orchestra gets louder we hear another original train sound effect, this time from the horns and trombones as together they slide upwards with their glissandi evoking an effective steam whistle warning of the train coming. Another realistic and original steam whistle effect comes from the woodwind section. The clarinet play a figure based on the chromatic scale, but using lip smears to slide from one note to the next.

As the train gathers steam, the main orchestral theme begins representing the train's progress through the countryside. This sinuous theme is typical of Brazilian folk music in the way that it uses descending melodic phrases in stepwise motion, along with some repeated notes. Rhythmically too it is typical with much use of syncopated (off beat) rhythms and the use of hemiola. A term used when three beats are performed in the time of two (or two beats are performed in the time of three). As in many Brazilian folk songs the syncopated rhythms are contrasted with a steady rhythmic pulse. Throughout this passage the South American percussion plays the same steady beat whilst the cellos and double basses offset this by playing on the beat.

As the train journey approaches its end, the train decelerates; the piece slows down and gets quieter. This is achieved partly by means of a clever rhythmic device by the piccolo and cello parts. To create the effect of the locomotive slowing down, the note values become gradually longer; the piccolo part starts with semiquaver sextuplets, then straight semiquavers, then triplet quavers, straight quavers, and finally triplet crotchets. The cello follows a similar pattern this time moving from semiquavers to crotchets. The note values continue to become longer, the woodwind and brass players drop out, leaving a series of sustained string chords interspersed with a few remaining scrapes and rattles which gradually die away as the train slows to a halt. All the instruments become silent apart from a long low chord on the cellos and then, out of the blue, the piece ends with a huge crashing dissonant chord.

Of the five pieces discussed in this chapter, one, *Le Train Bleu*, refers to the train only in its title and another, Poulenc's 'En chemin de fer', makes little attempt beyond a sense of movement to imitate a train. Each of the remaining three, however, is a musical evocation of a train ride: the journey begins; the train accelerates; it travels along and then decelerates and comes to a halt. Both Honegger and Ibert emphasise the mechanical nature of the machine but Villa-Lobos adds another dimension by exploring the 'rhythmic encounter between train and landscape'. As George Revill writes

> the composer binds together the rhythms of the moving train…with melodies inflected by the folk music of the rural hinterland through which the train runs. Together these cross-currents weave together the physical progress of the train through the landscape and the social and cultural world it inhabits.[24]

Revill goes further when he writes that because Villa-Lobos 'was central to the awakening of a national cultural consciousness in Brazil …this music and his work can be interpreted more broadly as an attempt to bind the nation together culturally and politically.[25]

Endnotes

1. In 1912 the composer and counterpoint teacher André Gédalge started a private class in orchestration for his most gifted pupils.
2. Aaron Copland and Vivian Perlis, *Copland, Volume I: 1900–1942* (London: Faber & Faber, 1984): 56-7.
3. Paul Griffiths. *A Concise History of Modern Music. From Debussy to Boulez.* (London: Thames and Hudson, 1978).
4. BBC Third Programme talk, 4 February 1962, quoted in Roger Nichols, *The Harlequin Years* (London, 2002): 276.
5. Aaron Copland. *What to Listen for in Music.* (New York: McGraw Hill Book Company, 1939).
6. George Revill. *Railway* (London: Reaktion Books Ltd., 2013): 56.
7. Quoted in Scowcroft and elsewhere.
8. As the railway expert Philip Scowcroft writes of the train speed 'This is a slight exaggeration for 1923, as it was not until 1938 that "Mallard" touched 126 m.p.h., the all-time record for steam traction'.
9. Arthur Honegger, quoted in Roger Nichols. *The Harlequin Years: Music in Paris 1917-1929.* (London: Thames and Hudson, 2002): 233.
10. *New York Times*, November 1, 1924.
11. *The Illustrated London News*, 14 Feb 1925.
12. *Truth*, 4 Feb, 1925.
13. *Westminster Gazette*, 19 September, 1925.
14. Charles O'Connell. *Victor Book of the Symphony.* (New York: Simon and Schuster, 1935): 289
15. *New York Times* Nov 1, 1924.
16. Ibid.
17. *The Musical Times*, Vol. 64, No. 966 (Aug. 1, 1923): 571-574.
18. As cited in Roxanne C. 'Composers and the Ballets Russes - Convention, Innovation, and Evolution as seen through the Lesser Known Works'. (PhD thesis The University of Manchester, 2016).
19. Carlo Patrão. 'Listening to the City of Light: An Interview with Sound Recordist Des Coulam'. https://soundstudiesblog.com/tag/paris-metro-system
20. Michael Round. 'Bachianas Brasileiras' in Performance'. *Tempo*, no. 169. (Cambridge: Cambridge University Press, 1989): 35.
21. Gerard Béhague. 'Villa-Lobos, Heitor 1887 – 1959' in *Grove Music Online*. https://www-oxfordmusiconline.
22. Some musicologists believe that movements I, II and IV are arrangements of earlier works for cello and piano.
23. A Nieweg Chart. Villa-Lobos: "Bachianas Brasileiras". Editions as of January / 2016. http://www.orchestralibrary.com/Nieweg%20Charts/Villa%20Lobos%20BB%202016.pdf
24. Revill, *Railway*, 59.
25. Ibid.

12 Four pieces by Percy Grainger and Charles Ives

THE FOUR PIECES featured in this chapter were written towards the beginning of the twentieth century by two composers who eschewed convention. Neither of them followed the usual path of a professional composer and each of the pieces is unusual and highly individual in its own way. There are two pieces evoking a train journey – Percy Grainger's *Train Music* and *The Celestial Railroad* by Charles Ives along with two pieces which are centered on train stations. 'Arrival Platform Humlet' was described by Grainger as 'The sort of thing one hums to oneself' whilst waiting on a station platform whereas 'From Hanover Square North, at the End of a Tragic Day, the Voices of the People Again Arose' recalls Ives' experience at a station the day that the news broke of the sinking of the RMS Lusitania.

Percy Grainger (1882-1961)

Percy Grainger is best known for his light music, notably *Handel in the Strand*, *Shepherd's Hey*, *Molly on the Shore* and *Country Gardens* which became a huge hit as *In An English Country Garden*. During his lifetime he was also renowned as a virtuoso pianist. But there was much more to him than his reputation as a celebrity pianist and a composer of light music. He was also a pioneering folksong collector and arranger, and an avant-garde thinker and experimentalist who worked with electronic music and instruments as early as 1937.

He was born in Melbourne, Australia, and was educated at home by his mother Rose an unorthodox creative spirit and a dominant influence on his life. As a talented young pianist he studied music in Frankfurt and then moved to

London to further his career as a concert pianist. Whilst in England he set out to collect and record folk songs with the Edison Phonograph, one of the first folk song collectors to do so. At the outbreak of World War I, he migrated to America and eventually settled in White Plains, New York where he earnt a fortune as a pianist; during the 1920s he earnt the equivalent of £50,000 per week.

Grainger was very close to his mother, some thought abnormally so, and after her suicide in 1922 when she jumped from the 18th floor of a New York sky scraper, he gave up his performing career for a time. In 1928 he married his Swedish girlfriend Ella Ström during a Hollywood Bowl concert featuring the première of his cantata *To a Nordic Princess* (1927–8) and with an audience of around 20,000.

It would be an understatement to describe him as a colourful and unconventional character. He dressed in multi-coloured home-made outfits and, bursting with energy, had a habit of leaping about and jogging from one concert venue to the next with a heavy rucksack on his back. He was fluent in 11 languages including Icelandic and Russian, but substituted Italian and

Large Railway Painting, 1920. Laszlo Moholy Nagy (1895–1946)

© Según indicaciones de los Titulares/ Entidad de Gestión de los derechos. Procedencia Museo Nacional Thyssen-Bornemisza, Madrid

German musical terms with Anglo-Saxon ones such as 'louden' rather than 'crescendo', and 'feelingly' instead of 'expressivo'. A charismatic personality, Grainger counted the composers George Gershwin, Duke Ellington, Edvard Grieg, and Frederick Delius amongst his friends. He died in White Plains Hospital in 1961 aged 78. In the 1930s he had established the Grainger Museum at the University of Melbourne - a repository of items documenting his life, career and music. He requested that after his death his skeleton should be placed on display there. His request was denied.

His musical output is enormous with many works having been 'dished up', as Grainger put it, for different combinations of instruments which included such rarities as the solovox[1] and theremin[2]. He also invented some of his own including the Butterfly Piano which had microtonal tuning producing a gliding sound. He composed in a broad spectrum of styles from highly experimental works such as *Free Music no. 2* for six theremins, to popular pieces such as his wind band setting of *Lincolnshire Posy* which he described as a 'bunch of musical wildflowers'. One of his preoccupations was his vision of 'free music' and he believed that for music to be completely 'free' it needed to be released from the 'tyranny of the performer'. To this end he worked towards the elimination of human intervention in performance by developing several 'free music' machines including the Kangaroo-Pouch Tone-Tool.[3]

Train Music

The vast majority of his compositions are for piano and much of it is very difficult to play. Grainger had very large hands and his piano music often uses a wide span. About 20 of the piano pieces are 'fragments' of projected series giving only the main theme of the work. *Train Music* is one such fragment. It exists in two versions - some sketches for large orchestra lasting less than five minutes, and a simplified piano version which lasts not much longer than 30 seconds. The pieces are rarely performed but there are several recordings available.

Grainger's mother accompanied her son on an extended European tour in the summer of 1900 and *Train Music* is his response to their travel in 'a very jerky train going from Genova to San Remo' (roughly 150 miles). The Genova-Ventimiglia railway was completed as a single track line in 1872. It is a scenic route often running close to the sea in the Liguria region of Italy. He started work on the piece when he was 18 and then played with the idea over the years making further jottings on the long railway journeys that he was partial to where he loved to spread his manuscripts out and compose.

Train Music is characterised by frequent changes of time signature and big percussive chords. This style of music was at odds with the prevailing romanticism of the time. With what Grainger referred to as its 'irregular barring', and its predominantly rhythmic texture and dissonant chords, it had more in common with Stravinsky's ground-breaking orchestral piece *Rite of Spring* (1913). Overall the effect is one of 'an alarming sense of rhythmic dislocation'.[4] According to Grainger-expert Bryan Fairfax,

> *Train Music* was born of a desire for a music to be created not only out of the sounds and rhythms of a moving train, but of its smell, heat, material, invincible 'onward rushing' (Grainger's words), changes of scene, climate and weather en route, heroic non-complaining in the service of mankind, and its consummate weight when at rest. A glorious totality; an essence transmuted into sound.[5]

The orchestral version is for 150 players, an orchestra of gargantuan proportions, and the equivalent of two symphony orchestras. There are parts for 100 strings and a massive woodwind section of 38 players rather than the usual number of eight players. They are mostly double reeds (eight oboes, four cor Anglais, six bassoons and two double bassoons) resulting in a strong nasal sound. The orchestral version was never realizable because of the large number of instruments needed but the fragment of the score was completed with a reduced orchestration in 1976 by Eldon Rathburn and can be found in the Grainger Archives.

'Arrival Platform Humlet'

'Arrival Platform Humlet' is the first movement of Grainger's four-movement suite *In a Nutshell*. It was put together from various pieces written at different times.

1. Arrival Platform Humlet
2. Gay but wistful
3. Pastoral
4. The Gum-suckers March

The movement 'Arrival Platform Humlet' exists in several versions - for solo piano, two pianos, solo viola or 'massed violas', along with a version for piano and orchestra with multiple tuned and untuned percussion instruments. The viola version is described as being for 'middle fiddle' owing to Grainger's aversion to the use of Italian terms in music. In Grainger's own words written at the top of the score, the piece was 'begun in Liverpool Street and Victoria Stations, London (England) on February 2, 1908' and 'scored for orchestra, piano and Deagan percussion instruments[6] as the first movement of my suite "In a Nutshell."' The musicologist Wilfrid Mellers describes it as a 'very odd piece that only Grainger could have thought up'.[7] At the top of the score Grainger writes

> Awaiting the arrival of a belated train bringing one's sweetheart from foreign parts: great fun! The sort of thing one hums to oneself as an accompaniment to one's tramping feet as one happily, excitedly, paces up and down the arrival platform. The final swirl does not depict the incoming of the expected train. The humlet is not programme music in any sense. It is marching music composed in an exultant mood in a railway station, but does not portray the station itself, its contents, or any event.

At this time of his life the 'sweetheart from foreign parts' may well have been his then lover and later long-time friend, the Danish pianist Karen Holten (1879–1953).

Grainger would mark his scores with idiosyncratic directions for the performers. The opening of 'Arrival Platform Humlet' is marked 'with healthy and somewhat fierce "go"', 'fierce' being a characteristic quality 'much admired' by Grainger. The opening woodwind theme of the orchestral version is marked 'nasal, reedy and snarling', a tone quality that classical woodwind players are more likely to avoid. Performance directions instruct the performers to play 'more clingingly' then 'less clingingly' and finally 'very clingingly'. The piece is mainly in octaves and unisons with no definite key and no predetermined form, rather it has the effect of spontaneous humming with snippets of tune strung together as they occur to the hummer, with little repetition'. In Grainger's words

> There are next to no chords in this composition, it being conceived almost exclusively in single line (unaccompanied unison or octave). There are likewise no 'themes' (in the sense of often-repeated motives), as the movement from start to finish is just an unbroken stretch of constantly varied melody with very few repetitions of any of its phrases.

These elements of 'free music' risked the piece lacking direction. However, towards its end, the command 'louden lots' moves towards an exciting conclusion marked by two chords to be played very loudly 'All you can'.

Charles Ives (1874-1954)

Charles Ives is now regarded by many as the leading American composer of the early twentieth century. Like Grainger, he too was a maverick and had an extraordinary working life. Although he studied music at Yale, composed and held various positions as a church organist, Ives did not pursue a professional career in music. Instead, in order to provide for his family, he opted for the more reliable insurance industry where he was very successful. So successful that after 20 years his company became the largest insurance agency in the country and he was soon a millionaire. He led a sort of double life where he composed in his free time and managed to create an immense body of music, even though he was only composing at weekends and on holidays. Inevitably he created his music in isolation which resulted in a highly original body of work in a wide variety of styles, from tonal Romanticism to radical experimentation, often embracing different stylistic characteristics within a single work, from atonality to polytonality, ragtime to ametrical rhythm. At the same time he drew on the music that he had grown up with such as popular American songs, parlour ballads, patriotic songs and hymn tunes, often quoting these within his pieces.

His works include symphonies and other orchestral works, amongst them the powerfully evocative piece *The unanswered question* and *Three places in New England* (also known as *Orchestral set no. 1*) where in one movement the conductor has to beat different times with each arm to represent the sound of overlapping bands marching, something which he had experienced when he was young. There are two piano sonatas, the second being the fearfully difficult *Concord sonata* where the movements are dedicated to important American writers – 'Emerson', 'Hawthorne', 'The Alcotts' and 'Thoreau'.

Ives worked in insurance for 30 years after which he devoted much of his time to revising his musical output and promoting its performance. He waited many years for recognition, partly because of the complexity of much of his music, partly because much of it had not been published, but after his death in

New York City in 1954, his reputation grew and his music eventually became a firm part of the concert repertoire.

'From Hanover Square North, at the End of a Tragic Day, the Voices of the People Again Arose'

Ives spent his summers in Redding, Connecticut and would commute every day to his job in New York over 60 miles away. It was a long haul, at least two hours between Redding and Grand Central Station. The journey started with a steam locomotive taking passengers down a single-track line and ended with the Third Avenue 'L' to Hanover Square, an elevated line in Manhattan.

'From Hanover Square North' is the third movement of the *Orchestral Set No. 2* which Ives assembled from material that he had composed between about 1909 and 1919 (although the first performance was not until 1967). The opening movement is 'An elegy to our forefathers' and the second movement is 'The Rockstrewn Hills join in the people's outdoor meeting'. The piece is scored for large orchestra which includes some instruments not normally found in this setting such as tuba, harpsichord, accordion, two pianos and two theremins. 'From Hanover Square North' adds a distant choir to the mix. It quotes several melodies that are well known in America: 'Massa's in de cold ground'; 'Ewing'; and 'My Old Kentucky home', but it is the hymn 'In the sweet bye and bye' (sometimes spelt as 'In the sweet by and by') that is central to the piece.

Ives recounted the origins of 'From Hanover Square North' in his Memos.

> We were living in an apartment at 27 West 11th Street. The morning paper on the breakfast table gave the news of the sinking of the Lusitania. I remember, going downtown to business, the people on the streets and on the elevated train had something in their faces that was not the usual something. Everybody who came into the office, whether they spoke about the disaster or not, showed a realization of seriously experiencing something. (That it meant war is what the faces said, if the tongues didn't.) Leaving the office and going uptown about 6 o'clock, I took the Third Avenue "L" at the Hanover Square Station. As I came on the platform, there was quite a crowd waiting for the trains, which had been blocked lower down, and while waiting there, a hand-organ, or hurdy gurdy was playing on a street below. Some workmen sitting on the side of the tracks began to whistle the tune, and others began to sing or hum the refrain. A workman with a shovel over his shoulder came on the platform and joined in the chorus, and the next man, a Wall Street banker with white spats and a cane, joined in it, and finally it seemed to me that everybody was singing this tune, and they didn't seem to be singing for fun, but as a natural outlet for what their feelings had been going through all day long. There was a feeling of dignity all through this. The hand-organ man seemed to sense this and wheeled the organ nearer the platform and kept it up fortissimo (and the chorus sounded out as though every man in New York must be joining in it). Then the first train came and everybody crowded in, and the song eventually died out, but the effect on the crowd still showed. Almost nobody talked-the people acted as though they might be coming out of a church service. In going uptown, occasionally little groups would start singing or humming the tune.[8]

> Now what was the tune? It wasn't a Broadway hit, it wasn't a musical comedy air, it wasn't a waltz tune or a dance tune or an opera tune or a classical tune, or a tune that all of them probably knew. It was (only) the refrain of an old Gospel Hymn that had stirred many people of past generations. It was nothing but-- "In the sweet bye and bye." It wasn't a tune written to be sold, or written by a professor of music--but by a man who was but giving out an experience.

This third movement is based on this, fundamentally, and comes from that "L" station. It has secondary themes and rhythms, but widely related, and its general makeup would reflect the sense of many people living, working, and occasionally going through the same deep experience, together...

The city noises of New York are represented by background ostinato. Over these dissonant sounds the orchestra takes up 'In the sweet by and by'. Both the verse and chorus are sung, but the hymn is fragmented, with strands of its tune trailing off with other melodies encroaching. The different layers create an increasingly dense and dissonant web as the tension builds up leading to a sudden release in a direct and heartfelt statement of the hymn's chorus. The song gradually trails away as the people leave and the piece ends.

The Celestial Railroad

The Celestial Railroad is a piece for solo piano, first performed in 1928. It is based on a short story of the same name by the American author Nathaniel Hawthorne. Hawthorne's story is in itself a parody of Bunyan's *Pilgrim's Progress* which tells of a Christian's spiritual journey through life. In Hawthorne's version a group of train passengers follow the route of Bunyan's Christian.

The story opens with a narrator who falls asleep and dreams that he is at a depot in the City of Destruction where he is befriended by Mr Smooth-it-away who takes him to board a waiting train bound for the Celestial City. The train sets off with its whistle blowing, stopping at the town of Vanity Fair, coming to rest at Beulah Land where the passengers are told to board a ferry which will take them across the River Jordan to the Celestial City. Once they are aboard the ferry the narrator sees the duplicitous Mr Smooth-it-away who has now changed back into his true devilish form. He realises that it has all been a hoax and jumps into the river to escape. This wakes him up, it was only a dream.

The Celestial Railroad is a complex piece which is very demanding to play; in fact Ives gives the performer alternative strategies for some of the most difficult passages. As in many of Ives' works the music includes references to hymn tunes and popular American songs. There are snatches of 'De Camptown Races' and 'Yankee Doodle', for example, but these are usually well-hidden and give only short bursts of the melody. Ives' work is a musical depiction of Hawthorne's story and, as such, follows the same course, even including some musical representations of train sounds. It opens with rapid flourishes swirling across the piano followed by a series of high bell-like chords suggesting the train whistle blowing as the passengers arrive at the depot. The train sets off and the chugging is represented by low cluster chords which get increasingly faster as the train accelerates.

This chapter has drawn together four differing pieces of railway music. In *Train music* and *The Celestial Railroad* both Grainger and Ives are drawn to the evocative sounds of a train journey. Both use percussive dissonant chords and a dense rhythmic texture to portray the movement of the train but Ives goes further in his imitation of train sounds whereas Grainger attempts to capture the smell, the heat, the onward rushing and the changes of scene in 'A glorious totality; an essence transmuted into sound'. Grainger's 'Arrival Platform Humlet' and Ives' 'From Hanover Square North' are both set on station platforms but they represent two very different experiences. Grainger's is a private affair where a lover hums to himself as he awaits his sweetheart, a single wandering line with no real themes or sense of direction. In contrast Ives represents a collective and intense experience where the dissonant tension of the music builds up throughout the movement until the crowd of passengers join together in a heartfelt rendition of the hymn tune 'In the sweet by and by'.

Endnotes

1 Grainger wrote for the solovox, an electronic sound-producing attachment which he added to the piano or Hammond organ.
2 A theremin is an electronic musical instrument controlled without physical contact by the player. It has two antennae, a vertical one which controls the pitch and a horizontal loop which controls the volume. Both are manipulated by the player's hands. It is named after the Russian Leon Theremin who invented it in 1920.
3 Kangaroo-Pouch Tone-Tool was developed by Grainger and the physicist Burnett Cross. It uses rolls of paper and a series of oscillators to produce a sound.
4 Wilfrid Mellers. *Percy Grainger*. (Oxford: Oxford University Press. 1992): 139.
5 Lewis Foreman, ed. *The Percy Grainger Companion*. (London: Thames Publishing. 1991): 71.
6 'Deagan percussion instruments' were manufactured by the Deagan company, founded in 1880. It manufactured glockenspiels and developed other instruments including the xylophone, vibraharp, aluminium chimes, aluminium harp, Swiss handbells, the marimba and tubular bells.
7 Mellers, *Percy Grainger*, 39.
8 Charles E Ives. *Charles E Ives. Memos*. (New York: W. W. Norton & Company, edited by John Kirkpatrick, 1972): 92.

13 Railway music after World War II

THE PIECES COVERED in this chapter are a disparate mix: they are by composers of different nationalities and are written for different instrumental and vocal forces. What the works do have in common is that, with one or two exceptions, they were written after World War II. The decades immediately following the war were politically turbulent and four of the featured composers - Ernst Krenek, Sergei Prokofiev, Kurt Weill and Hans Werner Henze – were émigrés. Part of the reason for them leaving their homeland was for their political convictions and the music of three of them had at some point been censured by the government of their homeland. Earlier chapters have described music written for the opening of lines and stations across Europe, but such celebratory pieces by British composers have been notably absent. A handful of these have finally appeared in recent decades and commemorative music by Michael Nyman, Paul Patterson and Graham Fitkin can be found in this chapter. The chapter features works in a range of styles for different combinations, voices, orchestras, and solo piano and it closes with a work for brass band – Philip Sparke's musical depiction of what is often thought to be the most exciting and exotic train service in the world – the *Orient Express*.

Music for voices

Ernst Krenek

The Austrian-American composer Ernst Krenek (1900 – 1991) loved trains, the travels they could take him on and the destinations they could take him to. As a child he loved to go to the railway yard of Wien Franz-Josefs-Bahnhof near his home in Vienna. When he started to compose, train travel became a recurring theme in his works from his smash hit opera *Jonny spielt auf* (Jonny strikes up) (1927), to his song cycle *The ballad of the railroads* (1944) and his choral piece *Santa Fe timetable* (1945).[1]

Jonny spielt auf (Jonny strikes up) (1926) – Ernst Krenek

Although Krenek's opera *Jonny strikes up* does not belong to the post-war period covered in this chapter it is worth a mention here, not least because a railway station is central to its plot and a train usually features as part of the set. Indeed one of the main characters is accidentally pushed off the platform into the path of an incoming train complete with dissonant warning sounds. The jazz-influenced work was premiered in Germany and was a huge success across Europe making an enormous amount of money. However, the work brought the opprobrium of the nascent Nazi party who saw the opera with its black musicians and shimmying dancers, as a symbol of the degenerate Weimar Republic. Subsequent intimidation led to Krenek's emigration, as a destitute refugee, when Hitler's troops invaded Europe. He eventually went to the USA where he taught at several universities and became an American citizen.

The ballad of the railroads, Op. 98 (1955) – Ernst Krenek

In his study of Krenek's life and music John Lincoln Stewart writes that 'railroads and rail travel fascinated him and had at times an emblematic, almost mystical, significance for him'.[2] In 1931 Krenek set the poems of Karl Kraus to music which included the line 'I dreamt of a traveling train'. He went on to compose a song cycle with a railroad theme, *The ballad of the railroads*, which uses his own lyrics. In his programme note to the piece, Krenek wrote

> ...many of my own emotional experiences and sentiments are reflected: the feelings of an uprooted man who entrusts himself to the trains, seeking a new home in foreign lands; hopes and fears engendered by the idea of travel; the agonies of separation and waiting; the life-long pull of the South and West; the arrival in this promised land on the shores of the Pacific, which is identified with the lost paradise of childhood.

Krenek's compositional style changed throughout his life and the music of the ten songs in this cycle is quite different from the jazz-influenced melodies of *Jonny strikes up*. Rather its style could be described as atonal serialism[3] where there is little sense of key. There is no mimicking of train sounds, but the piano helps to give a sense of travelling movement from the opening trills on the words 'Railroads, railroads, dinning in my ear' until the final song where the train 'pulls gently in' and the 'wheels stand still' to the accompaniment of sparse slow-moving chords.

Santa Fe timetable (1945) – Ernst Krenek

In *Santa Fe timetable*, an unaccompanied chorus intones the names of the stations between Albuquerque and Los Angeles on the Sante Fe railroad. Krenek had been inspired by *Liber Generationis Jesu Christi*, a choral piece by the Renaissance composer Josquin des Prez, which uses the first chapter of the Gospel of Saint Matthew for its text. In fact Krenek's beautiful piece has a religious feel to it. However this is offset by the words as the station stops are listed – 'Gallup, Gallup, Gallup', 'Hackberry, Yucca, Victorville' and the lively rhythms of 'Topoch, Topoch, Topoch' which, to Krenek, suggested the 'clicking of the wheels on the rails'.[4]

Winter Words (1954) by Benjamin Britten

In Benjamin Britten's song cycle *Winter Words* (1954) for high voice and piano, the train acts as a metaphor for the journey through life, from innocence to consciousness. The eight songs are based on poems by Thomas Hardy, and the second and seventh both tell tales of train travel. In 'Midnight on the Great Western' a 'journeying boy', ticket stuck in the band of his hat, is travelling third class knowing nothing of his destiny, 'Bewrapt past knowing to what he was going, Or whence he came.' The piano recalls the train's whistle and the movement of the engine as he hurtles through the night. The song opens with a train whistle sound which recurs several times throughout the journey, acting as a musical motif. In the opening bars the train is heard slowly coming closer in bumpy staccato quavers, the accompaniment starts to build up until jogging yet lilting rhythms of the train are established. The train rhythms drop in and out of the melancholy song, sometimes interrupted by the whistle motif, and sometimes by a momentary change of rhythm perhaps representing the changing of points. At the end of the song, the train is moving out of sight, it grinds to a halt and the whistle motif is played again, whistling 'from afar'.

The penultimate song in the cycle 'At the railway station, Upway' takes us back to the song about the journeying boy in the railway carriage. A young boy with a fiddle tries to cheer up a convict who is being taken away by a police constable. The themes of innocence and a journey into an unknown destination return but, in this dark and quasi-operatic song, there are no musical references to the train other than in the title.

'Train to Johannesburg' from *Lost in the Stars* (1949) – Kurt Weill and *Boulevard solitude* (1952) - Hans Werner Henze.

Mention should also be made of a couple of train-inspired movements from two dramatic vocal works written a few years apart. Both composers, Kurt Weill (1900-1950) and Hans Werner Henze (1926-2012), were born in Germany, both were politically active and both spent much of their life living elsewhere, Weill in America and Henze in Italy. Their compositional styles however were very different. Weill's most well-known music, such as 'The ballad of Mack the Knife', is founded on jazz and cabaret, whereas Henze's music, although having some jazz influence, is more cutting edge classical music. By 1939 Weill was settled in America when he composed and arranged music for one of the most popular attractions at the New York World's Fair. *Railroads on Parade* was a pageant telling of the effect of railroads on American life. It was remarkable for its use of locomotives and Pullman cars.[5] *Lost in the Stars* is a musical in two acts after Alan Paton's novel *Cry, the Beloved Country*. In 'Train to Johannesburg' Weill uses more conventional train imitations with repeated rhythms on the drum kit and train whistle sounds provided by the large woodwind section of clarinets and saxophones, and accordion.

Boulevard Solitude is Henze's first full-scale opera and is based on Abbé Prévost's novel *Manon Lescaut*. The opening of the opera is set in a railway station in a French town: Manon Lescaut is on her way to a boarding school accompanied by her brother. The scene opens with a sparse texture of percussion sounds, a station announcement is made and the percussion texture builds up into more regular repeated patterns.

Two pieces for orchestra

Both of these orchestral pieces were written for children: the first, *Winter*, was written for children to listen to; and the second, *Chat Moss*, was composed for children to perform. Sergei Prokofiev's *Winter Bonfire* is deliberately simple, whereas *Chat Moss*, by Peter Maxwell Davies, is surprisingly complex. Train journeys are central to both pieces.

Winter Bonfire, Op. 122 (1950) - Sergei Prokofiev

The Russian composer Sergei Prokofiev (1891 – 1953) left his homeland after the 1917 Revolution moving to the United States, then Germany, then Paris, making a successful living as a composer, pianist and conductor. In 1936 he returned home and enjoyed success there with compositions including *Peter and the Wolf* and *Romeo and Juliet*.

Winter Bonfire was commissioned by the Radio Committee in celebration of the approach of the thirtieth anniversary of the Pioneer movement, a youth organisation aiming to implement political and moral education and to develop socialist attitudes amongst children.[6] It is an eight-movement suite for narrator, choir and orchestra. It is set to a text by Samuil Marshak, a popular author of children's literature. The text describes the journey of the Pioneers who have travelled from city to countryside in order to visit a collective farm and it is read by the narrator before each movement begins. *Winter Bonfire* depicts the events of the children's outing in the snow, the departing train ride, snow falling behind the window, waltzing on the ice and the evening bonfire. It is probably the simplest work amongst Prokofiev's later works for children. This is evident in its use of straightforward harmony and repetition. Most of the opening movement does not leave the key of C major, and the final movement 'Return' repeats the locomotive themes and the sound of the trains' whistles of the opening movement. It is deliberately simple, partly to fulfil the contract which required that the music was to be accessible for children, but also to avoid any criticism from the Soviet authorities that the work was 'formalist' and did not conform to Socialist Realism. The definition of formalism was very vague, music would be criticised as formalist simply because it was not sufficiently Russian-like. Prokofiev once said with a note of satire that "Formalism is music that people don't understand at first hearing."[7]

The opening movement 'Departure' begins with the narration of a text of thirty six lines describing the arrival of the pioneers at the station early in the morning. The sound of a whistle is heard and they hurry to get on to the train. The orchestra evokes the sounds of an approaching train through such imitative effects as a regular beat on the timpani, a thundering roll on the snare drum, mechanical-sounding figurations in the strings, and blasts from the muted trumpets, all designed to recreate the sounds of the clatter of wheels and the blasts of a train whistle.

Prokofiev need not have worried about Soviet criticism and accusations of formalism; *Winter Bonfire* received an enthusiastic reception in the Soviet Union and was awarded a Stalin Prize.

Chat Moss (1993) – Peter Maxwell Davies

Peter Maxwell Davies (1934 – 2016) is widely regarded as one of the UK's leading composers. His compositions are extensive and in a variety of styles ranging from the radical experimentation of the music theatre piece *Eight Songs for a Mad King* (1978) to a series of ten large scale symphonies. Much of his early life was spent in Salford and Manchester, where he studied music.

From 1959 to 1962, Davies was director of music at Cirencester Grammar School, where the significance he attached to performance and composition, by pupils of all musical abilities, had a lasting impact on British music education. Through his teaching, Davies helped to show that children were far less resistant to the features of new music often considered displeasing by adults (and sometimes music teachers). He went on to write many pieces for school orchestras and community musicians. *Chat Moss* is an orchestral piece for school orchestra first performed by St. Edward's College, Liverpool. But this is not to imply that the music is easy; it is subtle and by no means straightforward. Davies set out to stretch young players and the piece is often played by professional orchestras. Furthermore it is much quoted in his Symphony No. 5 (1994).

Chat Moss is an area of Salford, close to Davies' childhood home in Leigh. Its underlying peat bog threatened the completion of the Liverpool and Manchester Railway because of the difficulties in creating a solid base. However, in 1829 the challenge was met by the engineer George Stephenson and his adviser Robert Stannard, an expert on marshland. Their ingenious solution was to 'float' the line on a bed of wood and rubble stone and in 1829, they succeeded in constructing a railway line through it. The line is still in use today.

The opening string melody is based on the plainsong *Haec Dies*. This quiet theme is developed across the woodwind and brass and then a more buoyant chugging rhythm is introduced underlying more variations. The music slows to a more sedate pace with a lazy oboe melody but then the pace picks up again, brass, timpani and crashing cymbals come to the forefront and the excitement builds as steam trains thunder by. Towards the end, the music falls to a hushed orchestral passage, the brass interject with a piercing chord reminiscent of a train whistle, and then comes to a quiet close.

Three pieces written to celebrate railway openings.

As we have seen in previous chapters, many orchestral pieces were written by well-known classical composers across Europe to celebrate the opening of new stations and railway lines in the nineteenth century. But, although broadside ballads often announced the arrival of new lines in the UK, there are no entries for celebratory pieces by British composers. It is hard to know why, although there may be a class element involved in the absence of music commissioned by British composers. Although the UK invented the railways, they upset the status quo, industrialised countryside and sped up the pace of life. So it could be that railways were not seen as a fitting subject for 'high' art. It appears that in other countries the train also takes a more central appearance in classic literature. Zola's novel *La Bête humaine* (1890), for example, is based upon the railway between Paris and Le Havre, and trains are a recurring motive throughout Tolstoy's *Anna Karenina* (1887) with several major plot points taking place on the Russian railways. However, there is no classic British novel where trains or railways are central to this degree other than some short stories. On the other hand trains and railways do feature prominently in British nineteenth century popular fiction such as crime novels and sensation novels.

Not until the twentieth century were there any railway pieces commissioned for the launching of new lines or trains.

MGV (Musique à Grand Vitesse) (1993) – Michael Nyman

MGV (*Musique à Grand Vitesse* - High-speed music) was commissioned by the Festival de Lille for the inauguration of the TGV North-European line and was first performed by the Michael Nyman Band and the Orchestre national de Lille on 26 September 1993. The *TGV Train à Grande Vitess* high-speed train is France's intercity high-speed rail service. Over the years the *TGV* has broken several speed records. The fastest single long-distance run was from Calais-Frethun to Marseille at a speed of 306 km/h (190 mph) for the inauguration of the *LG Méditerranée* in 2001.

The compositions of Michael Nyman (b. 1944) combine elements of classical music with hi-energy repetitions. His distinctive style is familiar from his many film scores, notably *The Draughtsman's Contract* and *The Piano*. Many pieces, including this one, are performed by the Michael Nyman Band, an unusual combination of saxophones, brass, strings and electric guitar. MGV is characterised by its layers of ostinati (repeated bars), different speeds and rhythms, and changing speeds. The recurrent rhythms are striking. The opening movement establishes cycles of 9, 11, or 13-beat rhythmic cycles which are heard against a regular rhythmic cycle of 8 creating interesting cross rhythms. The piece is relatively simple harmonically with chord sequences built mainly over C and E. The piece runs continuously, but it is divided into five sections, 1st region, 2nd region etc. each lasting about five minutes. Nyman describes these as 'five inter-connected journeys, each ending with a slow, mainly stepwise melody'. Only when we reach the end of the journey, the 'destination', do we hear the triumphant melody in its full glory replete with brass fanfares.

The Royal Eurostar Opus 76 (1994) – Paul Patterson

Paul Patterson (b. 1947) was commissioned to write *The Royal Eurostar* by European Passenger Services for the State Opening of Waterloo International and the inauguration of the Channel Tunnel in 1994. It was written for 19 brass and percussion players with optional organ. The 12-minute piece builds on a sustained D which corresponds to the hum of Eurostar engines. It incorporates both 'Rule Britannia' and the 'Marseillaise'. The 'Eurostar Fanfare' was originally conceived as the opening to *The Royal Eurostar*. The plan was for it to be played on the arrival of a train bringing the Queen from France.

However, it became a separate piece after it was decided that the Queen would travel to, not from, France. 'Paris Fanfare' celebrated the arrival of the Queen at Waterloo, and was performed for the inauguration of the London-Paris service on 16th November 1994. On the same day the 'Brussels Fanfare' was used for the launch of the London-Brussels Eurostar service.

Track to track: The Athlon (2012) – Graham Fitkin with words by Glyn Maxwell.

Track to Track was commissioned by the London Chamber Orchestra to celebrate the start of the 'Javelin' shuttle service between St Pancras and the Olympic site at Stratford ready for the 2012 London Olympics. It was composed as part of an education project where children came up with rhythmic motives for Fitkin to use in the piece as well as ideas for the text which were brought in the by the writer Glyn Maxwell. *Track to track* is scored for ensemble and string orchestra. The music was designed to be broadcast on the train with different music for the six-minute outward and return journeys, some of it palindromic. The piece opens quietly and reflectively with sustained chords. Descending glissandi (slides) follow and we hear the words 'I rose so very early that day, so early that day'. As the day begins and the journey commences a rhythmic ostinato is heard, building up as the train approaches the Olympic site. The ostinato continues as the train makes its return journey becoming livelier and jazzier, but, in the manner of a palindrome, returning to the sustained chords and downward glissandi of the opening and the words 'I rose that day so very early'.

The final two pieces in this chapter are unconnected and very different. The first is a short complex piano piece by a cutting-edge contemporary composer, and the second is a much lighter crowd-pleaser, a wind band piece written in a popular vein.

Freightrain Bruise (1972 rev 1980) by Michael Finnissy

Freightrain Bruise is a short jazz-infused piano piece by the British composer Michael Finnissy (b. 1946). Like many of his works it is rhythmically complex and intricate in texture. The scores are often labyrinthine in appearance. For some years Finnissy worked with dancers and dance companies and he wrote several pieces that were written either to accompany dancers or to capture on paper his spontaneous improvisations. *Freightrain Bruise* was written in collaboration with Charlotte Holtzermann, an American dance student in London at The Place, and it is dedicated to her. The piece is influenced by the music of jazz musicians such as Art Tatum, Errol Garner, Thelonious Monk and it is marked 'Lightly and quietly, but with a raunchy swing' at the top of the score. At the same time the jazz elements are filtered through modernist language; it is written in a chromatic fashion without a strong sense of key, and is fragmented with intermittent silences. Irregular groups of notes are played against each other, not just three against two but five against four and six against five. Finnissy has described these elements as 'bruises on its surface, places where the ripe fruit of jazz hit the cold floor of late twentieth-century angst.'[8] Nevertheless it is music to dance to, there is a constant sense of movement as the freight train trundles along, chugging away until it comes to an abrupt halt and the piece closes.

Orient Express (1992) – Philip Sparke

In 1883 the *Orient Express*, the epitome of comfort and luxury, crossed the whole of Europe entering places little known to most Western Europeans. The inaugural train left Paris on October 4 scheduled to reach Constantinople in three and a half days. The engineer behind it was Belgian Georges Nagelmackers, founder of the *Compagnie Internationale des Wagons-Lits*. In a series of letters about his journey, a correspondent for *The Times* described the train as having a level of 'comfort and facility hitherto unknown'.[9] It had an elegant dining room serving gourmet food, ladies boudoirs, a library, Turkish carpets, upholstered beds, and showers with running hot and cold water. The route covered 1800 miles and crossed seven borders. As well as the small number of luxury carriages, there were also several basic ones for third-class passengers travelling from one Eastern European country to another. The *Orient Express* soon became popular leading to the inauguration of a variety of other routes, all with variants on the name. A second line opened in 1918, the *Simplon Orient Express* which used the Simplon Tunnel between Switzerland and Italy travelling via Milan, Venice and Trieste and in the 1930s, yet another route started on the *Arlberg Orient Express*.[10] A London connection was added in between the wars with a Calais ferry and rail link to Paris. However, by 1962 partly owing to the spread of motor car ownership, only the *Simplon Orient Express* was left. The route was shortened several times until in 2009 with the opening of the high-speed line between Paris and Strasbourg, it ceased to operate at all.

Orient Express is a wind band piece and was commissioned by the Tokyo Kosei Wind Orchestra. In his programme note, the composer Philip Sparke writes

> The piece describes a journey on the famous luxury train which runs from London to Venice. Starting among the hubbub of London's Victoria Station, the music subsides as the guard blows his whistle and the enormous engine starts rolling. At last the train is under way, hurtling across the European countryside. A slower central section gives the passengers time to think about home but the relentless journey continues, eventually slowing to a halt as the train reaches its destination.[11]

Venice Simplon Orient Express poster
Retro AdArchives / Alamy Stock Photo

In the passages where the music imitates the sounds of the train, *Orient Express* is notable both for the instruments it is scored for and the way the instruments are used. In Sparke's quest to replicate rolling engine setting off, the usual wind band instrumentation of woodwind, brass and percussion is supplemented by a guard's whistle, a steam whistle and sandpaper blocks. A piercing blast from the guard's whistle and the first jittering movements of the train are heard in the percussion; snare drum played with alternate sticks and brushes, with what gives the impression of randomly-placed sandpaper blocks. All this is underpinned by intermittent crushed chords on muted trombones. The silences between the notes gets shorter until a repeated pattern is established across the percussion, brass and lower woodwind, all the time getting louder and faster as the train accelerates. The high woodwind enters with a long dissonant sliding chord which fades out for the first main theme of the journey to enter. The Orient Express is on its way.

Endnotes

1. https://www.krenek.at/en/forum
2. John Lincoln Stewart. *Ernst Krenek. The man and his music* (Los Angeles: University of California Press, 1992): 252.
3. Serialism is a system devised by the composer Schoenberg to replace tonality (keys) in his music. It is sometimes known as the twelve tone system.
4. Stewart, *Krenek*, 249.
5. The music was not published, but in 1992 David Drew composed a concert suite, *Trains Bound for Glory*, using some of Weill's folksong arrangements taken from it.
6. Joseph I. Zajda. *Education in the USSR*. (Oxford: Pergamon, 1980): 82.
7. Boris Schwarz. *Music and Musical Life in Soviet Russia, 1917-1970*. (New York: W W Norton, 1973): 115.
8. https://openaccess.city.ac.uk/id/eprint/22296/21/Finnissy%20-%20The%20Piano%20Music%20%284%29.pdf
9. *The Times*, November 2, 1883.
10. Christian Wolmar. *A Short History of Trains*. (London: Dorling Kindersley Ltd., 2019): 196.
11. https://musicum.net/orient.pdf

14 A more experimental approach

EACH OF THE six pieces in this chapter is ground breaking in its own way. The first, Pierre Schaeffer's 'Etude aux chemins de fer', represents one of the earliest experiments in electroacoustic music whereas in the second, *U.S. Highball*, Harry Partch uses a newly invented scale and newly invented instruments. 'Etude aux chemins de fer' is arguably the first piece of music to use sampling, and sampling is central to Steve Reich's politically motivated piece for string quartet and tape, *Different Trains*. In the remaining three pieces, *The End of the Line (A Brief Encounter)* by Peter Wiegold and Dominic Power, Gavin Bryars' *The Stopping Train* and John Cage's *Il treno*, the railway is not only the subject of the composition, it also serves as the performance venue itself. In five of the six pieces, the connection between the music and the train lies, to some extent, in the recreation of the rhythmic sounds of the railway. Cage's piece is different. Instead it could be argued that it is more about the atmosphere of the train journey, acting as an invitation to be aware of the sounds that are all around us. Random sounds are taken from the train and its environs and everyday sounds are juxtaposed with manufactured ones; the piece is unpredictable and full of possibilities, generating new auditory experiences and having the potential to transform the way we think about both music and listening.

'Etude aux chemins de fer' (1948) - Pierre Schaeffer

It is ironic that a central aim of the composer of one of the most groundbreaking railway pieces was that 'the train must be forgotten'. Thus wrote the French composer, acoustician and electronics engineer Pierre Schaeffer (1910-

1995) of his pioneering piece 'Etude aux chemins de fer', the first of his *Cinq études de bruits (Five studies of noises)*. These were the first pieces of *musique concrete*, a form of electroacoustic music where natural everyday sounds, rather than musical instruments, are recorded and then distorted by simple techniques of editing, reversal and speed changing. Schaeffer began his career at France's public broadcaster Radiodiffusion Française (RDF) in 1936. He persuaded the RDF to give him permission to begin experiments in music technology, researching into noise, and it was here, at the RDF Studio d'Essai that the five *études* were created.

'Etude aux chemins de fer' is a collage of sampled train sounds recorded at Batignolles station in Paris. The short work opens with the sound of a distorted train whistle followed by a succession of loops of the six engines puffing and chugging, slowed down, speeded up, superimposed, changing in volume, and all punctuated by transformed whistle sounds. One of the most significant developments was the idea of making loops of sound. Schaeffer used phonograph discs (he later started to use tape) and the loops were made by a studio technique referred to as the 'locked groove', where the disc was scratched so that it would play a sound in a perpetual loop.

Musique concrete demanded new ways of thinking and writing, as well as listening. Schaeffer wished to break the link between a sound and its cause. He argued that listening without reference shifted the listener's perception to other purely sonic attributes and he coined the term 'acousmatic' to describe a sound when its source is hidden.

> I hope that one day there will come together an audience that prefers the theoretically less rewarding sequences, where the train must be forgotten and only sequences of sound color, changes of time, and the secret life of percussion instruments are heard.[1]

The works were first broadcast on 5 October 1948 with the title 'Concert de bruits' and were met with some hostility. Nevertheless, the importance of these pieces cannot be denied; they presented a new way of making music, manipulating and transforming sounds electronically, which has had a lasting influence. Schaeffer's ideas were central to electronic music and the work of such composers as Iannis Xenakis, Pierre Boulez and many others who followed. But Schaeffer's influence was to reach well beyond the field of contemporary art music. One of the most significant developments was the idea of making loops of sound and in doing so he effectively invented sampling – the mainstay of today's hip-hop and electronic dance music (EDM).

Espace (Escape) (1989) – Francis Dhomont

In the same year that Schaeffer composed 'Etude aux chemins de fer', the French composer Francis Dhomont (b. 1926) was also experimenting with electroacoustic music. Some forty one years later he used train sounds in his piece *Espace* (Escape) a poetic reverie on wandering. *Espace* (1989) is an acousmatic piece involving shifts of space, place and atmosphere. The sounds of trains, environments in railway stations and congregations of people are manipulated in the studio using different processors and then diffused through multiple speakers in performance.

U. S. Highball (A musical account of a transcontinental hobo trip) (1943, revised 1958) - Harry Partch

Harry Partch (1901 – 1974), the American composer, instrument maker and performer, was born in California but grew up hearing music from many cultures as he travelled around with his missionary parents. His musical inspiration and materials came from outside the European tradition and because of this he was ignored by most musical institutions. Partch rejected Western scales and invented his own which was made up of 43 notes in the octave rather than 12. He went on to devote his time implementing this system, having to create new instruments which could be tuned to this new scale and then training performers to play them. In his lifetime Partch was often considered to be simply eccentric, but since his death he has come to be recognized as one of the most innovative American composers of the twentieth century.

Much of his life was spent living on the margins of society. In 1935 at the height of the Great Depression he began a transient existence in the western states, jumping trains, sometimes living in boxcars in train yards, picking up work and occasionally obtaining grants from organizations en route. For several years he kept a journal, a small notepad that he used to transcribe 'fragments of conversations, remarks, writings on the sides of boxcars, signs in havens for derelicts, hitchhikers' inscriptions, names of stations, thoughts.'[2] These notes became the basis for *U.S. Highball*, a piece for speaker and Partch's instruments. In 1943 he started work on the first draft using the motley assortment of instruments he had built at that point. These included the Adapted Guitar (a guitar with new frets to accommodate his tuning system), the Chromelodeon (a re-tuned foot-pump harmonium), and the Kithara (his own version of the ancient Greek stringed instrument but now with 72 strings). Partch took the comments he had recorded in his notebook and turned them into melodies which matched the inflection and declamation of American English.

In *Genesis of a Music*, the book where the composer laid out his compositional and aesthetic treatise, Partch wrote that U.S. Highball was the 'most creative piece of work I have ever done'. The second version, which Partch preferred, was completed in 1958. New instruments were added which he had created over the previous ten years resulting in a more percussive sound. The instruments included a set of marimba-based instruments; a smaller version of the 72 string Kithara (Surrogate Kithara), and a collection of tuned artillery shells and Pyrex water containers (the Spoils of War). The work includes quotes from the hoboes such as "Wait for the next drag, there will be lots of empties on it. Too cold to ride outside this weather", "You'll get killed on that oil tank. There's an empty back here", "Ain't got no matches, ain't got no tobacco, ain't got no money. Is that blanket big enough for two?" "There she jumps again. That engineer don't know how to drive this train" and "Hey, don't sleep with your head against the end of the car! You'll get your neck broke when she jerks!" With the small intervals in his new system of tuning applied to both voice and instruments, sliding steam whistle sounds and the occasional percussive rhythm, *U. S. Highball* has a surreal atmosphere, but it is oddly reminiscent of a train journey.

Partch's composition uses fragments of speech transformed into melodies. Reich uses a similar technique in *Different Trains* by taking the pitches of small samples of speech and then transcribing them in musical notation to be played by a string quartet.

Different Trains (1988) - Steve Reich

Steve Reich was born in New York in 1936. Having completed a degree in philosophy he went on to study composition with Luciano Berio and Darius Milhaud amongst others. Following this more conventional classical music education, he went on to study Balinese gamelan, African drumming and traditional forms of the chanting of Hebrew scriptures subsequently incorporating these non-Western elements into his music. In the 1960s, along with Terry Riley and Philip Glass, Reich discovered the tape-based techniques of phasing and looping. Some pieces included recordings of fragments of speech (*Come Out* and *It's Gonna Rain*, for example), others, such as *Clapping Music*, were based on fragments of musical material which were then 'phased'.[3] His music has influenced generations of pop, jazz and classical musicians.

Different Trains (1988) is in three movements played without a pause.

1. America – Before the war
2. Europe – During the war
3. After the war.

Reich writes in his programme note

> The idea for the piece came from my childhood. When I was one year old my parents separated. My mother moved to Los Angeles and my father stayed in New York. Since they arranged divided custody, I travelled back and forth by train frequently between New York and Los Angeles from 1939 to 1942 accompanied by my governess. While the trips were exciting and romantic at the time I now look back and think that, if I had been in Europe during this period, as a Jew I would have had to ride very different trains. With this in mind I wanted to make a piece that would accurately reflect the whole situation.

In order to prepare the tape I did the following:

1. Record my governess Virginia, then in her seventies, reminiscing about our train trips together.

2. Record a retired Pullman porter, Lawrence Davis, then in his eighties, who used to ride lines between New York and Los Angeles, reminiscing about his life.

3. Collect recordings of Holocaust survivors Rachella, Paul and Rachel, all about my age and then living in America—speaking of their experiences.

4. Collect recorded American and European train sounds of the '30s and '40s.[4]

In the first movement, Reich's governess and the Pullman porter Lawrence Davis, reminisce about train travel in the USA while American train sounds are heard in the background. European train sounds and sirens are heard in the second movement. Three Holocaust survivors speak about their experiences in Europe during the war, including travelling to concentration camps by train. In the third movement the Holocaust survivors talk about the years immediately following World War II, alongside recordings of Davis and Virginia and a return to the American train sounds from the first movement.

Reich selected small samples of speech that used fairly clear pitches and then transcribed them as accurately as possible in musical notation. This was to be played by the strings, so in effect, the strings imitate the melodies of the speech. In Reich's words 'The piece thus presents both a documentary and a musical reality and begins a new musical direction'. Reich used Casio FZ-I and FZ-IoM samplers to manipulate and play the fragments of speech. In the same way the train sounds, sirens and bells are sampled and edited and incorporated into the scheme of the music. So, for example, the repeated semiquavers of the string writing are coupled to the clatter of trains.[5]

The British composer and musicologist Christopher Fox observes that the 46 spoken phrases are the most remarkable feature of the work.

> ...through them Reich is attempting nothing less than a brief history of perhaps the most appallingly systematic onslaught, in this or any other century, by a government on the lives of millions of people. By focussing on the personal histories of a few individuals he is able to emphasize the inhumanity of the Nazis' invasion of so many people's lives; the juxtaposition of the two Americans with their European contemporaries establishes the contrast between normality and the Europeans' experiences. Thus when the Pullman porter, Lawrence Davis, says in the third movement, 'But today, they're all gone', he is recalling the luxurious transcontinental trains on which he worked; however, for the listener, these words can also become an elegy for the millions of people who died between 1933 and 1945.

In the final movement Reich addresses what Fox describes as 'one of the fundamental questions posed by the Holocaust: how is it possible that the same music can be enjoyed by both oppressed and oppressor?' Rachella, the Holocaust survivor describes how 'There was one girl, who had a beautiful voice, and they loved to listen to the singing, the Germans, and when she stopped singing they said, "More, more" and they applauded'.⁶

The world premiere of *Different Trains* for string quartet and pre-recorded tape took place in 1988. Following its UK premiere, Paul Griffiths, writing for *The Times*, described it as a 'quite astonishing new piece'.

> It is entirely new in its voice-instrument fusions, and in the further fusions with the recorded train noises, notably sirens and whistles. It is new too in its rapid, railroad-paced changes of tonality, effected stunningly in the middle section by glissandos along lines laid out by the sirens. In this section there is also the dizzying impression of a whole network of musical traffic behind the main train, with chains of repeating figures being switched through points, appearing from and vanishing into silence.⁷

In the following three pieces the railway is central, not only for its sounds, but also for its performance venue. In his piece 'From Hanover Square North', Charles Ives' recalls his experience when crowds waiting on the station platform broke into the spontaneous singing of a hymn on the day that the news broke of the sinking of the RMS Lusitania (see pages 157 - 59). The first piece in this group is resonant of this memory, a bittersweet event taking place in a station amongst large crowds.

The End of the Line (A Brief Encounter) – Peter Wiegold composer and Dominic Power libretto

On July 4, 2009, 150 musicians and eight solo singers from the Royal Northern College of Music took part in a musical installation at Piccadilly Station in Manchester. Under the direction of the composer Peter Wiegold, and with a libretto by Dominic Power, the students mingled amongst an audience of commuters, clubbers and late-night travellers to perform a new piece, *The End of the Line (A Brief Encounter)*. The piece paid homage to the 1945 film *Brief Encounter*, famous for its romantic scenes set in a railway station (filmed on location at Carnforth), not least the heart-rending scene when the couple part for the last time. The eight singers, dressed in 1940s fashion, made up four couples who were parting and saying their farewells. The station clock has stopped and they are frozen in a moment of time as their memories flash by.

All in all there were two string orchestras, two brass bands, percussion, a group of clarinets and saxophones playing repeated rhythmic figures reminiscent of a train, and a bank of harps adding a romantic note. The groups of musicians were sited at various parts of the station - the singers on podiums, woodwind and brass on the gallery, some amongst the audience - several with their own conductors dressed as railway guards. Layers of sound, exploiting the sonic potential of the vast space of the station, produced, at times, a magnificent cacophony'. As the work came to an end and the music died away, one train sounded a long note on its horn as the singers disappeared, one by one, down the long platform.

The Stopping Train (2018) - Gavin Bryars and Blake Morrison

In 2016 the Yorkshire Festival and Sound UK commissioned Gavin Bryars and Blake Morrison to create an audio work to be listened to on the train journey from Goole (Bryars' birthplace) to Hull. Morrison reads his own poems and members of the Gavin Bryars Ensemble play the music. While Morrison's poetry describes the journey, musing on the area's history and geography, the music underscores the texts. The ten tracks are timed to coincide with each of the five stops along the route in both directions and are composed for headphone-wearing passengers as they gaze out of a train window. Stuart Jeffries took the journey and describes it thus.

> The words are accompanied by the viola, cello and bass of the Gavin Bryars Ensemble, with an electric guitar mournfully wailing like a train whistle as we roll beneath grey skies through a captivatingly lugubrious landscape of low-lying sheep fields and windfarms…says he thinks of his native land as a melancholy place. "Despite or perhaps because of that, it has its fascinations," he says.[8]

Alla Ricerca del Silenzio Perduto (In Search of Lost Silence) 1978 – John Cage

On June 26, 1978 a three-day event began in Bologna Centrale Station with the title *Alla Ricerca del Silenzio Perduto*. A stopping train filled with passengers, musicians and electronic music equipment embarked on its journey through the north-eastern Italian region of Emilia-Romagna making stops at designated towns. Passengers were greeted at the station by bands and dancers from the surrounding area, speakers relayed sounds from the local environment, and cassette recorders played unexpected sounds in random corners.

The shorter version of the title was *Il Treno di John Cage*. The piece had been inspired by John Cage (1912 – 1992): experimental composer, inventor of the prepared piano, champion of chance music and pioneer of the happening, the man who claimed that anything and everything was music. In a Seattle lecture Cage once said

> Everything we do is music. Wherever we are, what we hear is mostly noise. When we ignore it, it disturbs us. When we listen to it, we find it fascinating. The sound of a truck at 50 miles an hour. Static between the stations. Rain. We want to capture and control these sounds, to use them, not as sound effects but as musical instruments.[9]

He challenges our preconceptions of what music is and dares us to imagine what it might be. In his most (in)famous piece, 4'33" (1952), the so-called 'silent piece', he asks the performer(s) to do nothing for the specified time, inviting the audience to be aware of the beauty of the sounds that are all around us.[10] Similarly, in capturing and controlling the sounds of the train and its environs in *Alla Ricerca del Silenzio Perduto* we are invited to listen in a different way.

Some have compared the way Cage uses the train as an instrument to his 'prepared piano', an instrument he invented where, by placing a variety of objects between the strings, it is transformed into something sounding more like a large percussion ensemble. The event did indeed take a good deal of preparation. The train was equipped with internal and external speakers. Sixteen microphones were placed in and around the wagons to gather the sounds of the train, the squeaks, the rattles and rhythms of the railway. Once collected the sounds would be amplified and mixed by Cage and his assistants and then sent through the speakers – one set per car. Video-cameras and screens were installed in the wagons, so that people could see what was happening in the other cars. As well as seven passenger cars, the train was equipped with a car for sound mixing to take place and one for the electricity supply. Throughout the journey musicians would walk around the train playing music and at each stop previously recorded sounds would be played through the external speakers.

Martin Friedman, the curator of the Walker Art Center in Minneapolis was aboard the train. Reminiscing about the event he writes that Cage had agreed to come to Bologna where Tito Gotti, the Bolognese impresario who had put forward the idea of the event, introduced him to the composers Walter Marchetti and Juan Hidalgo and then set about choosing the right train. As part of the audition process they decided to look for simple cars, not every train

John Cage aboard Il Treno di John Cage, 1978
© Photograph Corinto Marianelli

would qualify. The plan was for the passengers to 'experience the juxtapositions of man-made and mechanically generated sound, unexpected sounds could be coaxed from parts of the auditory mix; nothing was sacred.' Two cars without compartments were used to house the chamber ensembles. At each of the stops the train was greeted by local people, bands, folk dance groups, and other ensembles creating a carnival-like atmosphere. Friedman writes 'As the picturesque ... panorama of tile-roofed villages and verdant farmland rolled by, passengers could switch channels to hear amplified train noises or pre-recorded music from the towns and villages where the train stopped'. Some of the musicians strolled round the cars with performers ranging from mime artists to vocalists. The first stop was at the little town of Riola where they were greeted by crowds and musician: the local police band played marches; choral music and folk songs were sung by the alpine infantry corps in feathered caps; and a village orchestra played romantic waltzes. News of the musical event spread up and down the lines the train would travel over in its three excursions with many more joining in the festivities.[11]

How did the event end? In its preview of the happening, the Italian newspaper *l'Unità* reported that

> During the return trip there will be no stops. The trip will be accompanied by the sound of the train only. Finally, when the train comes to a complete stop, we will hear the sounds of the station at night: empty and calm; and we will find the Lost Silence.[12]

Endnotes

1. Pierre Schaeffer. *In Search of a Concrete Music*, trans. Christine North and John Dack. (Berkeley and Los Angeles: University of California Press, 2012): 14.
2. Andrew S Grenade. "U. S. Highball (A Musical Account of a Transcontinental Hobo Trip). Harry Partch; Gate 5 Ensemble (1958)", 2004. https://www.loc.gov/static/programs/national-recording-preservation-board/documents/USHighball.pdf, 1.
3. Phase shifting (or phasing) is where one part repeats constantly and another gradually shifts out of phase with it.
4. Reproduced by kind permission of Boosey & Hawkes.
5. Christopher Fox. Steve Reich's 'Different Trains' in *Tempo*, Mar., 1990, New Series, No. 172 (Mar., 1990): 6.
6. Fox, 'Different Trains', 8.
7. *The Times*, Nov. 3, 1988.
8. *The Guardian*, 14 April, 2018.
9. John Cage. *Silence. Lectures and writings by John Cage.* (Middletown: Wesleyan University Press, 1961).
10. Barry Russell. *Everything we do is music: Cross-curricular experiments in sound based on the music of John Cage.* (London: Peters Edition, 2016).
11. https://walkerart.org/magazine/il-treno-di-john-cage-happening-on-wheels
12. *L'Unità*, May 23, 1978.

Railway music in North America in the nineteenth and early twentieth century

15 The coming of the railroads to North America: work songs, hoboes, gospel music and the blues

THE RAILROAD IS a central motif in American work songs, folk music, gospel, blues and jazz. The titles, lyrics and sounds of the songs are filled with images of the railroad: it offered a promise of escape and access to employment; it transported people, culture and ideas in ways previously unknown; and travel by train altered people's perception of time. The following chapters outline some of the ways in which the railroad featured in American popular music.

Transport in the United States was slow and difficult until the coming of the railroads (the name 'railroads' was adopted early on). The roads were poor and although there were canals, these froze in the winter and the system did not cover the whole country. Baltimore was without a canal system and in 1827 the Baltimore and Ohio Railroad was incorporated, bringing locomotives and track from England. Soon, however, Americans began to build their own locomotives and in 1830 *Tom Thumb* was built for use on the Baltimore line after another early company, the South Carolina Canal and Rail Road Company, had the first locomotive built for commercial use in the United States – the *Best Friend of Charleston*. Canada, also largely dependent on canals invested in trains and founded the Champlain and Saint Lawrence Railroad in 1832.

Rapid technological advances in steam locomotive design in the late 1830s led to a huge expansion in railway building. This development accelerated in the 1840s and 1850s and the growth of the US railroad exploded. By the 1860s trains had largely replaced canals as transport. This was to have a profound effect on the USA. In 1800 a journey from New York City to Detroit took four weeks, whilst the same journey in 1860 could be completed overnight. There was still a recognised need for a transcontinental railroad to traverse from east to west. The Union Pacific and Central Pacific railroad companies were the first to realise this goal. In the 1860s a labour shortage threatened the railroad's completion; local people were unwilling to work on the railroad finding mining and gold-digging more lucrative. In 1864 around 15,000 Chinese workers were shipped in to help build the Central Pacific railroad. In fact, according to the American historian Peter Liebhold, all the workers on the railroad were 'other', 'On the west, there were Chinese workers, out east were Irish and Mormon workers were in the centre. All these groups are outside the classical American mainstream.'[1] The work was hard; the railroad was built entirely by manual labourers who had to face dangerous work conditions – accidental explosions, snow and rock avalanches – and hundreds died as a result. The workers lived in camps that were moved along with the railroad. These ungoverned temporary villages of tents and shacks were known as 'hells on wheels'.

1869 marked the completion of the first transcontinental railroad when the Union Pacific and Central Pacific were joined together in Utah. This event was celebrated by a ceremony known as the Golden Spike (also known as the Last Spike) where a final 7.6-carat gold spike was driven. By 1916 the USA had the biggest railway network in the world covering 250,000 miles. Partly as a result of this success, the nation thrived and the railways bound the vast country together.

This chapter outlines some of the early development of train-inspired American songs focusing on work songs and the building of the railroads, moving on to the folk songs, country music, and blues that document the wandering hobo lifestyle, its dangers and consequences.

Work songs

The coming of the railway in the States created an enormous demand for labour to construct the railroads and tunnels. Slave trading had been abolished from the beginning of the nineteenth century onwards and by the middle of the century there were large numbers of free black people in the USA, however they were still denied many of the rights enjoyed by the white population. During this period black labourers predominated in the railroad building workforce working alongside white, often Scottish and Irish, manual workers. One of the earliest types of black American music, and a forerunner of the blues, was the work song. Although slaves had been discouraged from many types of music making, work songs were encouraged because they were sung rhythmically in time with the task being done and this actually helped the work. At the same time, worksongs helped to relieve the boredom of usually tedious jobs. Many work songs used call and response where one member of the gang would lead the song with the others coming in after him.

With their strong rhythms, work songs helped to coordinate labour on the railroad. When hammering in spikes to hold down the rails and ties for example, workers would swing their hammers in a full circle, hitting the spikes, one after the other, without missing. The most effective way to accomplish this was to set up a rhythm, traditionally provided by a work song. Scott Reynolds Nelson (2008) describes how a hammer song coordinated the movements of the workers.

> The hammer man swung a sledgehammer down onto the chisel. The shaker shifted the drill between blows to improve the drill's bite…The breaks between the lines of the hammer songs coordinated the complex movements of drilling. The hammer came down at the end of the line…Sometimes it was a hammer man who sang, telling his partner with his rhythm and lyrics when the next blow would come. Other times a third man would sing for hammer man and shaker. Between the blows a shaker would work his magic, either rocking or rolling.[2]

Sometimes the words of the work songs had their roots in British folk music or broadside ballads; sometimes they were improvised. They often followed an AAB form where the first and second lines were the same – a tradition which was later followed in the blues. As Ted Gioia writes in his book *Work Songs* (2006) 'No tool plays a more significant and complex role in the surviving body of work songs than the hammer'.[3] The word 'hammer' took on whole new meanings as a symbol of the worker's strength, determination or virility. Many blues singers and folk singers have recorded versions of a work song called 'Take this hammer' a defiant song in which a prisoner contemplates escape.

Take this hammer
Take this old hammer, take it to the captain,
Take this old hammer, man take it to the captain,
Tell him I'm goin', tell him I'm gone.

If he asks you, was I runnin',
If he asks you, was I runnin',
Tell him I was flyin', man tell him I was flyin'.

Stripes but no Stars, rail road workers, Asheville, North Carolina, USA, c.1892
Buncombe County Special Collections, Pack Memorial Public Library, Asheville, North Carolina, USA

Prisons were a source of cheap labour for building and maintaining railroad tracks until the 1940s. Work songs continued to be sung, particularly by work gangs in the southern US gaols. During the 1930s the American Library of Congress sent out folklorists to collect field recordings of American folk music. John A Lomax and his son Alan toured US gaols in the southern states where many blues singers were incarcerated. 'Take this hammer' was recorded at State Prison Farm, Raiford, Florida in 1939. The prisoners included the later celebrated blues singer Huddie Ledbetter (Leadbelly) who had been jailed for assault and whose repertoire included work songs. Lomax was so impressed by Leadbelly's singing and 12-string guitar playing that on his release he hired him to travel round state penitentiaries acting as a driver, mechanic and assistant for more recording sessions. Another version, which Robert Winslow Gordon heard when collecting songs on his US gaol visits in the 1920s, makes reference to John Henry and includes the following lines.

This ole hammer-huh!
Killed John Henry – huh!
Can't kill me, baby – huh!
Can't kill me!
My ole hammer-huh!
Shine like silver – huh!
Shina like gold!
Aint no hammer-huh!
In this whole mountain – huh!
Shina like mine, baby – huh!
Shina like mine!

Work songs were, of course, sung a cappella but some of them developed into songs accompanied by guitar. In 1928 the blues singer Mississippi John Hurt recorded 'Spike driver blues' which includes the following lines

Take this hammer and carry it to my captain
Tell him I'm gone, tell him I'm gone, tell him I'm gone

This is the hammer that killed John Henry
But it won't kill me, but it won't kill me, but it won't kill me

It a long ways to East Colorado
Honey to my home, honey to my home, that where I'm going

Ted Gioia argues that 'John Henry' stands in a class by itself' and that 'any assessment of work songs in American music and life leads inevitably to the consideration of John Henry – half historical figure and half legend – and his compelling story.'[4] For this reason John Henry will be dealt with in more detail later (see pages 207 - 210).

Hoboes

10 railway songs about hoboes

SONG TITLE	EARLY RECORDINGS AND FIRST DATE OF PUBLICATION
Because he was only a tramp	Published and performed by Hamilton 1875–1880
Hobo Bill / Hobo Bill's last ride	Published by Waldo O'Neal, 1929 Jimmie Rodgers, Gene Autry, Martha Copeland
Hobo blues	Published by Bernard Besman and John Lee Hooker Peg Leg Howell, Dr. Isaiah Ross, Cliff Carlisle
Hobo's lullabye	Published (1934) and performed by Goebel Reeves, later by Arlo Guthrie
Hobo's meditation	Published by Jimmie Rodgers Joe Glazer, Jimmie Rodgers, Hank Snow, Ernest Tubb.
Railroad tramp	Walter Morris, Dock Boggs
The dying hobo	First published in 1909 in *Railroad Man's Magazine*. First recorded by Dick Burnett and Leonard Rutherford in 1926.
The railroad boomer	Published and performed by Carson Robison, 1929 Gene Autry, Carter Family, Goebel Reeves
The Wabash Cannonball	Roy Acuff, The Carter Family, Hugh Cross (1929)
Waiting for a train (Wild and reckless hobo)	Jimmie Rodgers (1929), Gene Autry, Ernest Tubb

The word 'hobo' is generally used to refer to impoverished American migrant workers who travelled around the country taking free train rides in search of work. The term came into use in the late nineteenth century. Some of the earliest black hoboes were freed slaves who worked at various jobs in the South. Many jobs were seasonal, others included construction jobs and railroad track maintenance, with many of the workers hopping freight trains to get to the job site. By the 1920s, hoboes, sometimes referred to as 'tramps', had become a ubiquitous presence anywhere near railroad tracks.[5] The numbers of hoboes increased dramatically during the Great Depression of the 1930s with estimates ranging from two to four million. Paul Oliver (1990) in his study of meaning in the blues writes

Far too many had no prospects whatsoever and toured aimlessly, with the police ready to arrest them for vagrancy or for failing to pay their fares. Living from day to day, skipping aboard the freight trains as they rattled slowly across high trestle bridges, clambering on to the passenger trains as they gathered speed on leaving the station behind, they let the great locomotives carry them to distant cities.[6]

It was a dangerous business; freight hopping involved riding under, on top of, or in the boxcars carrying the freight, risking injury, sometimes being beaten off by hostile train crews and pursued by violent security officials. The penalty could be prison or the chain gang. Riding on a freight train was dangerous in itself and there was always the possibility of arrest. Two of the earliest songs depicting this sorry life are serious pieces of social commentary. 'Because he was only a tramp' is a poignant country fiddle song by Wyzee Hamilton that dates from the 1870s.

Because he was only a tramp

I'm a broken man without credit or cash
My clothes are all tattered and torn
Not a friend have I gotten in this wide world alone
I wish I had never been born.

Now it happened right here on the AGS run
A man was all wearied and worn
Along by the tracks sat an empty boxcar
He crawled in and closed up the door.

Now he hadn't rode far in this empty boxcar
'Til the brakeman came round with his lamp
He was thrown from the car and was killed on the bar
Because he was only a tramp.

Come now hear all my good people
Don't call every poor man a tramp
He was starin' in the face by starvation you know
That will bring down a man to a tramp.

Hoboes lived in 'jungles' set up by the side of the track. As the blues expert Paul Oliver writes, these were 'primitive shack towns made from scrap metal, wood, and cardboard and the packing cases that hoboes had tipped off the trains'.[7] Oliver goes on to write that 'The railroad held few illusions for the blues singer…As transient, as tramp, as wanderer, or worker he sang from experience rather than sentiment'.[8]

Hobo blues

In his study of black hoboes and their songs, Paul Garon (2006) writes that in this way 'Blues singing hoboes documented their railroad experience in song' and that the lyrics themselves are 'material evidence of history'.[9] Although John Lee Hooker's 'Hobo Blues' is not identical to a 'material document like a government form…it does record a mood and an attitude, and as such, is a legitimate document'. The song was written at a time of economic disaster in the Dust Bowl and massive unemployment of farmers and labourers in the South where many of them became 'hoboes, migrant laborers or travellers to the North'.[10] Blues singer and guitarist John Lee Hooker (1917 – 2001),[11] from a sharecropping family, ran away from his home in Mississippi at the age of 14, reportedly never seeing his mother again. He travelled around working in various cities until he settled in Detroit in 1948. In 'Hobo blues' Hooker cries out to the Lord, telling of riding the freight trains a long way from home and away from his mother. Albert Murray was the first to write at length about the connection between the blues and locomotive onomatopoeia in his book *Stomping the Blues* (1976). John Lee Hooker's rhythmic boogie-blues style shows what Murray describes as the 'influence of the old smoke-chugging railroad-train engine'.[12]

Hobo blues as sung by John Lee Hooker
When I first start to hoboing, hoboing, hoboing
I took a freight train to be my friend, oh Lord
You know I hoboed, hoboed, hoboed, hoboed
Hoboed a long, long way from home, oh Lord

You know my mother, she followed me that mornin', me that mornin', boys
She followed me down to the yard, oh Lord
She said, "my son he's gone, he's gone, he's gone, he's gone
Yes, he's gone in a poor some wear, oh Lord"

Yes, I left my dear old mother, dear old mother, dear old mother
I left my honor, need a crime, oh Lord
You know I hoboed, hoboed, hoboed, hoboed
Hoboed a long, long way from home, oh Lord

Big Rock Candy Mountain

'Big Rock Candy Mountain' is a country song that tells of a utopia for hoboes, free of the stresses and concerns of everyday life. It has a convoluted history and exists in several versions, some clean, others far less so.[13] Harry "Haywire Mac" McClintock (1882-1957), who claimed to be the song's original composer and recorded it in 1928, created two very different versions of the song. Both were superseded by later versions, notably Burl Ives's 1945 recording. Ives version is closely related to McClintock's clean version and includes the following refrain.

I'll show you the bees in the cigarette trees,
And the soda water fountain
And the lemonade springs where the blue bird sings.
In the Big Rock Candy Mountains.

McClintock's other version relates to the pre-Depression hobo culture where older, experienced hoboes would recruit young runaway boys. These relationships often began with the older hobo, the 'jocker' describing the wonders of life on the road. In exchange for the promise of safety and the promised joys, the younger hobo, the 'punk' would agree to perform various tasks ranging from menial labour and begging to sexual acts.[14] This version makes several references to life on the rails for hoboes.

In the Big Rock Candy Mountains all the jails are made of tin,
And you can walk right out again as soon as you are in,
Why, the brakemen have to tip their hats
And the railroad bulls are blind,

In the Big Rock Candy Mountains there's a land that's fair and bright
Where the handouts grow on bushes and you sleep out every night,
Where the boxcars all are empty and the sun shines every day.

Hobo Bill's last ride

Riding on the eastbound freight train speeding through the night
Hobo Bill the railroad bum was fighting for his life
The sadness of his eyes revealed the torture of his soul
He raised a weak and weary hand to brush away the coal

No warm lights flickered round him no blankets there to fold
Nothing but the howling wind the driving rain so cold
When he heard a whistle blowing in a dreamy kind of way
The hobo seemed contented for he smile there where he lay

Riding on the eastbound freight train speeding through the night
Hobo Bill the railroad bum was fighting for his life
The sadness of his eyes revealed the torture of his soul
He raised a weak and weary hand to brush away the coal

No warm lights flickered round him no blankets there to fold
Nothing but the howling wind the driving rain so cold
When he heard a whistle blowing in a dreamy kind of way
The hobo seemed contented for he smile there where he lay

Outside the rain was falling on that lonely boxcar door
But the little form of Hobo Bill lay still upon the floor
As the train sped through the darkness and the raging storm outside
No one knew that Hobo Bill was taking his last ride

It was early in the morning when they raised the hobo's head
The smile still lingered on his face but Hobo Bill was dead
There was no mother's longing to soothe his weary soul
For he was just a railroad bum who died out in the cold.

It is ironic that one of the best-known hobo songs was written by a professional songwriter, Waldo O'Neal, rather than a homeless tramp. However, 'Hobo Bill's last ride' was written for a one-time hobo, Jimmie Rodgers (1897-1933), the Singing Brakeman, who had spent many years on the railroads both hoboing and working. Born in Mississippi, the son of a railroad gang foreman on the Mobile and Ohio line, he worked in various capacities on the railroads until tuberculosis forced him from that work. He had always longed for a career as a musician, having learned to play guitar and banjo, and from 1925 this became his focus, although he still had stints or railroad work all over the Southern States. He picked up musical ideas from fellow railroad men and from listening to his vast collection of phonograph records. He evolved his own country music style which had elements of the blues he had learnt from black musicians. Norm Cohen (2010) in his landmark study of American railroad folksong *Long Steel Rail: The Railroad in American Folksong* describes Rodgers' style as 'a form of white blues – a pale imitation of the classic city blues'.[15] Rodgers was also a yodeller sometimes interspersing yodelling between the verses of his songs. The lyrics often welded together 'the images of cowboy and railroad wanderer'.[16] In 1927 he made a trial recording and its success showed record companies that there was a large audience for this type of music which they referred to as 'hillbilly'; his musical style did much to set the pattern for country music. Not surprisingly the Singing Brakeman made much use of railroad imagery in songs such as 'Waiting for a train', 'Train whistle blues', 'The brakeman's blues', 'Southern Cannonball' and 'The hobo's meditation'. Rodgers' recording of 'Hobo Bill's last ride' is a finger picking country song interspersed with yodelling, whereas another of his big hits, 'Waiting for a train' is performed in a more bluesy, swing vein. Both songs describe the unemployment and hardship of a hobo's lonely life.

Waiting for a train

All around the water tanks
Waiting for a train
A thousand miles away from home
Sleeping in the rain
I walked up to a brakeman
To give him a line of talk
He says if you've got money
I'll see that you don't walk
I haven't got a nickel
Not a penny can I show
Get off, get off, you railroad bum
He slammed the boxcar door

He put me off in Texas
A state I dearly love
The wide open spaces all around me
The moon and the stars up above
Nobody seems to want me
Or lend me a helping hand
I'm on my way from Frisco
I'm going back to Dixieland
Oh, my pocketbook is empty
And my heart is full of pain
I'm a thousand miles away from home
Just waiting for a train.

Rodgers continued to be hugely successful with subsequent records, but tuberculosis led to his premature death in 1933. Many country stars followed in his footsteps including Gene Autry, Cliff Carlisle, Hank Snow and Wilf Carter. For the next 20 years after Rodgers' death most major country singers had one or more hits in his railroad-inspired vein. These included Boxcar Willie. One might be forgiven for thinking that Boxcar Willie had been a hobo given his stage name, his unkempt stage persona (wearing train driver's overalls and a battered hat), his repertoire of railroad songs and his train whistle impressions. In fact he had actually spent 22 years of his working life as a flight engineer in the US Air Force.

Robert Johnson

The Mississippi blues singer Robert Johnson is often referred to as a hobo. He certainly spent much of his short life travelling from place to place, an inveterate wanderer he moved from town to town across North America playing on street corners and in juke joints to raise money, often picking up women along the way. He is believed to have been born in 1911 and died a violent death in 1938, allegedly after being poisoned by a jealous husband. In this short but turbulent life he had little commercial success and recorded only 29 songs, accompanying himself on slide guitar. However his work has since been hugely influential and many of his songs have become blues classics such as 'Rambling on my mind', 'Love in vain' and 'Walking blues', all of which refer to trains.

Rambling on my mind

I got ramblin', I got ramblin' on my mind (twice)
Hate to leave my baby but you treat me so unkind
I got mean things, I got mean things all on my mind (twice)
Hate to leave you here, babe, but you treat me so unkind
Runnin' down to the station, catch the first mail train I see (twice)
I got the blues about Miss So-and-So and the child got the blues about me.
And I'm leavin' this mornin' with my arm' fold' up and cryin' (twice)
I hate to leave my baby but she treats me so unkind.
I got mean things, I've got mean things on my mind (twice)
I got to leave my baby, well, she treats me so unkind.

For Johnson the train was often a means of escape. As the musicologist Dick Weissman writes '…examining Robert Johnson's lyrics brings to life a person who was a restless dissatisfied wanderer, always seeking out the next woman, and seemingly never, or rarely, believing that she would offer anything more than temporary companionship and an outlet for the singer's sexual needs of the moment'. Like several of his other songs, 'Rambling on my mind' focuses on his distrust of women and his need to get away.[17] The slide guitar work is typical of the Delta blues, there is a walking bass rhythm and short instrumental breaks (solos) after each line of the lyrics, a sort of call and response, almost as though the guitarist is in conversation with the singer.

Walking blues

I woke up this mornin', feelin' round for my shoes
Know 'bout it I got these, old walkin' blues
Woke up this mornin', feelin' round for my shoes
But you know 'bout 'it I, got these old walkin' blues.

Well I leave this morn' I have to ride the blind,
I feel mistreated and I don't mind dyin'
Leavin' this morn', I have to ride a blind
Babe, I been mistreated, baby, and I don't mind dyin'.

'Walking blues' reflects the hobo lifestyle; itinerant musicians were sometimes known as 'walking musicians'. It was originally written and recorded in 1930 by another itinerant musician, the Delta blues singer Son House. Both Robert Johnson and Muddy Waters later recorded their own versions. Johnson's recording uses various slide guitar techniques, thumbed strumming on the lower strings and fingerpicking on the treble strings as well as featuring some well-placed falsetto. Riding the blinds was the dangerous hobo practice of riding between cars on a moving train, out of sight of the train crew or police. On a passenger train this area between the cars was usually covered with canvas or leather. On freight cars, hoboes sometimes rode holding onto the ladder running up to the top of the car, this equally dangerous act was also referred to as riding the blinds.

Love in vain

'Love in vain', is a sad romantic song of unrequited love that uses a departing train as a metaphor for a lost love. The singer follows his love to the station and cries as she leaves. The acoustic guitar accompaniment is sparse with the longing vocal line very much in the forefront. In the last wordless verse the lyrics disappear and are replaced by a lonesome cry.

Love in vain

And I followed her to the station, with a suitcase in my hand (twice)
Well, it's hard to tell, it's hard to tell
When all your love's in vain
All my love's in vain

When the train rolled up to the station, I looked her in the eye (twice)
Well, I was lonesome, I felt so lonesome
And I could not help but cry
All my love's in vain

When the train, it left the station, with two lights on behind (twice)
Well, the blue light was my blues
And the red light was my mind
All my love's in vain

Johnson recorded his songs in the 1930s, and since then hundreds of versions have been made by blues and rock bands across the world. He has had a great influence on the rock musicians of today notably Eric Clapton and the Rolling Stones whose version of 'Love in vain' has sold over a million copies.

Women with the train blues

The 1930s was a time of massive unemployment in the Southern states of the USA. Many of those who lived there travelled north in search of work. Most travelled by train, a means of transport which began to take on a new symbolism in a migration that took them far away from home and split their families apart. The following three blues songs were all recorded in the 1930s and they are all sung by women, the singer/guitarist Memphis Minnie and the classic blues singers Bessie Smith and Lucille Bogan. Two of the songs, 'I hate that train called the M&O' and 'Chickasaw train blues', blame the train for taking their man away and the third, 'Dixie Flyer blues', welcomes the train as a means of getting back home to Dixieland, the nickname for the Southern United States.

I hate that train called the M&O as sung by Lucille Bogan
I hate that train that they all call the M&O (twice)
It took my baby away and he ain't comin' back to me no more.

When he was leavin', could not hear nothin' but that whistle blow (twice)
And the man at the throttle swore: "He ain't comin' back no more".

He had his head in the window, that man was watchin' those drive wheels roll (twice)
Said, "I'm goin' away baby and I ain't comin' back no more"

The M&O refers to the Mobile and Ohio Railroad which ran from Alabama to Ohio. Like most other female blues singers, Alabama born Lucille Bogan (1897-1948), sometimes known as Bessie Jackson, wrote most of her songs herself. She was known for her bawdy lyrics which were often about adultery, prostitution and lesbianism; her songs were often so outspoken and dirty that they had to be censored by the record companies. Not in this song however, instead in 'I hate that train…' Bogan directs her venom at the train that is taking her man away. Paul Oliver writes of the way in which the train was often personified in the blues with the singer addressing complaints to it and the train becoming a 'scapegoat for faults that might well have been laid at the door of the singer' or the 'subject for the blameworthy when events could not be explained'.[18]

Chickasaw Train blues (low down dirty thing) by Memphis Minnie

I'm goin' tell everybody, what that Chickasaw has done done for me (twice)
She done stole my man away, and blow that doggone smoke on me
She's a low down dirty dog.

I ain't no woman, like to ride that Chickasaw (twice)
Because everywhere she stop, she's stealing some woman's good man, oh
She's a low down dirty dog.

I told the depot this mornin', I don't think he treats me right (twice)
He done sold my man a ticket, and I know that Chickasaw leavin' town tonight
He's a low down dirty dog.

I walk down a railroad track, that Chickasaw even wouldn't let me ride the blind (twice)
She's a low down dirty dog.

Mmm-mmm, Chickasaw don't pay no woman no mind (twice)
And she stop pickin' up men, all up and down the line.

Memphis Minnie (1897-1973) was an accomplished blues guitarist, vocalist, and songwriter. So skilled was she as a guitarist that she is reputed to have played in a contest against Big Bill Broonzy and won.[19] A common theme in the blues as sung by women puts the focus on cheating men at the same time as asserting the woman's strength and independence in seeking revenge. In this song Memphis Minnie personifies the Chickasaw train as a 'low down dirty train', 'picking up men, all up and down the line'. The lyrics are full of sexual imagery. As Oliver points out, in many blues songs, sexual intercourse is figuratively known as 'riding'.[20] This is the way that Memphis Minnie uses it in the phrase 'like to ride that Chickasaw'. Her prowess as a guitarist is evident in this song, not least in the intricate finger picking.

Dixie Flyer blues as sung by Bessie Smith
Hold that engine, catch me mama, get on board
Hold that engine, catch me mama, get on board
'Cause my home ain't here, it's a long way down the road.

On that choo-choo, mama gonna find a berth
On that choo-choo, mama gonna find a berth
Home into Dixieland, it's the grandest place on earth.

Dixie Flyer, come on and let your driver roam
Dixie Flyer, come on and let your driver roam
Wouldn't say enough to save nobody's doggone soul.

Blow your whistle, tell them mama's comin' too
Blow your whistle, tell them mama's comin' too
Take it up a little bit, 'cause I'm feelin' mighty blue.

Here's my ticket, take it please, conductor man
Here's my ticket, take it please, conductor man
Homin' to my mammy, way down into Dixieland.

The widely renowned Bessie Smith (1894-1937) was the most popular female blues, vaudeville and jazz singer of the 1920s and 1930s and became the highest-paid black entertainer of the day. She had a long and successful recording career. The Empress of the Blues was an influential performer, as Paul Oliver writes 'Her broad phrasing, fine intonation, blue-note inflections and wide expressive range made hers the measure of jazz-blues singing in the 1920s.'[21] All of these can be heard in 'Dixie Flyer blues'. The *Dixie Flyer* was a passenger train that operated from 1892 to 1965 via the 'Dixie Route' from Chicago and St. Louis via Evansville, Nashville, and Atlanta to Florida. In Smith's song, recorded in New York in 1925, the train is personified as a friend that is taking the singer home 'to my mammy' in Dixieland the 'greatest place on earth', and a cure for the blues. Bessie Smith is accompanied by Buster Bailey on clarinet, Charlie Green on trombone and Fred Longshaw piano. Unlike the previous blues songs described here, 'Dixie Flyer blues' imitates the train in the accompaniment. The song opens with the sound of a train whistle and bell as Smith intones the words 'Hold that train'. The train sets off with rhythmic engine imitations provided by James T Wilson shuffling gradually accelerating sandpaper blocks. At the words 'Dixie Flyer, come on and let your driver roam', the clarinet plays descending slides and the words 'Blow your whistle' are answered by toots on the train whistle. Above all this, Bessie Smith sings a heartfelt blues.

The development of blues music overlapped with the evolution of gospel music. Although gospel music was short on train sounds, its lyrics were permeated with railroad imagery. Part of the reason for this was that in the days of slavery, railroads and trains began to symbolise the road to freedom.

Spirituals and gospel music

Before emancipation, slaves needed a written bond from their owner if they were to take a train journey. Such a journey would, for most slaves be only a dream, hence its allure found its way into spirituals and gospel music which often used train metaphors using such phrases as 'Getting on board the Gospel Train'. The early to mid-nineteenth century saw the formation of the clandestine Underground Railroad, a loosely knit organisation whose purpose was to help slaves in the South escape to the North, often Canada. It used a network of secret routes and safe houses and was assisted by abolitionists and others sympathetic to the cause. The Underground Railroad was often referred to as the 'freedom train' or 'Gospel train', which headed towards 'Heaven' or the 'Promised Land'. Religion offered the slaves hope, the hope that one day they would be led to freedom in the 'promised land'. During the early nineteenth century massive Protestant 'camp meetings' attracted thousands of people, both white and black. These were extended religious services, taking place over several days and including a great deal of singing. It was during this period that the spiritual emerged as a distinct type of religious song.[22] These were sung by groups of people, often using call and response; a call by a lead singer would be answered by the congregation. In the places where blacks worshipped separately they would adopt their preferred musical styles.

African vocal style included slides and slurs, whistles, yodels, and changes in rhythm and types of sound, and these were incorporated. Singers would be encouraged to improvise, and foot-stamping and clapping with up-beat tempos were often used. In this way the impassioned style known as gospel music evolved.[23]

'This train'

This train is bound for glory, this train.
This train is bound for glory, this train.
This train is bound for glory,
Don't carry nothing but the righteous and the holy.
This train is bound for glory, this train.
This train don't carry no gamblers, this train (twice)
This train don't carry no gamblers,
Liars, thieves, nor big shot ramblers,
This train is bound for glory, this train.

This train don't carry no liars, this train; (twice)
This train don't carry no liars,
She's streamlined and a midnight flyer,
This train don't carry no liars, this train.
This train don't carry no smokers, this train (twice);
This train don't carry no smokers,
Two bit liars, small time jokers,
This train don't carry no smokers, this train.

The origins of the spiritual 'This train' are unknown, but it started to become popular in the 1920s and first became a hit in the 1930s when it was recorded by Sister Rosetta Tharpe. Since then it has remained in the repertoire with many recordings and different versions of the lyrics. In 1934 it was included in John and Alan Lomax's anthology *American Ballads and Folk Songs* and later became popular during the folk revival in the 1950s and 1960s. The most successful versions include those by Woody Guthrie, Peter, Paul and Mary and more recently Mumford and Sons.

'Life's railway to Heaven' as sung by Johnny Cash

Life is like a mountain railway
With an engineer that's brave
We must make the run successful
From the cradle to the grave
Heed the curves and watch the tunnels
Never falter, never fail
Keep your hand upon the throttle
And your eye upon the rail.

Blessed Saviour, Thou will guide us
Till we reach that blissful shore
Where the angels wait to join us
In that train forevermore
Oh, Blessed Saviour, Thou will guide us
Till we reach that blissful shore
Where the angels wait to join us
In God's grace forevermore.

'Life's railway to Heaven' has been a popular hymn for many years and has been recorded in several different styles, not just gospel but bluegrass, country and western and folk. It was first published in 1890 with the copyright going to M E Abbey (lyrics) and Charles D Tillman (music). However, Norm Cohen believes that it was modelled on an older poem by the songwriter William Shakespeare Hays in 1886.[24] As Cohen observes

The theme of the spiritual railway – the allegorical representation of our earthly sojourn as a railroad trip, the ultimate destination of which is heaven – was exploited in many nineteenth-century poems and songs. In extended metaphors such songs designated the path of piety, with the implicit warning that to stray from the route delineated by the steel rails of righteousness would lead elsewhere but heaven.[25]

In some religious songs the train was headed very much in the opposite direction if the passengers had not lived a virtuous life. The lyrics to 'Little black train' as recorded by the Carter Family and Woody Guthrie, for example, act as a warning.

'Little black train'

There's a little black train a-comin'
Comin' down the track
You gotta ride that little black train,
But it ain't a gonna bring you back.
You may be a bar-room gambler
And cheat your way through life
You can't cheat that little black train
Or beat this final ride.

You silken bar-room ladies,
Dressed in your worldly pride
You've gotta ride that little black train
That's comin' in the night.
Your million dollar fortune,
Your mansion glittering white
You can't take it with you
When the train moves in the night.

The songs in this chapter all provide some kind of social commentary. The railroads offered the way out of a desperate situation, as symbolised in the Underground Railroad, and in the railroad imagery of spirituals and gospel music. Work songs helped to provide a record of the hardship in building the American railroads and other songs described here tell of the dangerous hobo lifestyle. Many songs of this period are about travelling home to the Southern States, documenting a time when, because of massive employment, thousands had to emigrate far away from home. In this way, the popular and religious music documented here makes an important contribution to American social history.

Endnotes

1. *The Guardian*, 18 July, 2019.
2. Scott Reynolds Nelson. *Steel Drivin' Man: John Henry, the Untold Story of an American Legend.* (Oxford: Oxford University Press, 2006).
3. Ted Gioia. *Work Songs.* (Durham and London: Duke University Press, 2006): 152.
4. Gioia, *Work Songs*, 160.
5. Graham Raulerson. 'Hoboes, Rubbish, and "The Big Rock Candy Mountain"', (American Music, Vol. 31, No. 4, 2013: 420-449): 424.
6. Paul Oliver. *Blues Fell This Morning: Meaning in the Blues.* (Cambridge: Cambridge University Press, 1960): 62.
7. Ibid.
8. Oliver, *Blues Fell*, 68.
9. Paul Garon. *What's the Use of Walking If There's a Freight Train Going Your Way?: Black Hoboes & Their Songs.* (Chicago: Charles H Kerr Publishing Company, 2006):7.
10. Ibid.
11. Different birth dates can be found for John Lee Hooker.
12. Albert Murray. Stomping the Blues. Minneapolis: University of Minnesota Press, 1976.
13. For a full account and discussion read Graham Raulerson. 'Hoboes, Rubbish, and "The Big Rock Candy Mountain"', American Music, Vol. 31, No. 4, 2013: 420-449.
14. Raulerson, "The Big Rock Candy Mountain", 427.
15. Cohen, *Long steel rail*, 30.
16. Ibid.
17. Dick Weissman. *Blues. The Basics.* (London: Routledge, 2005): 68.
18. Oliver, *Blues Fell*, 66.
19. Weissman, *Blues*, 80.
20. Oliver, *Blues Fell*, 107.
21. Paul Oliver, 'Bessie Smith' in *Grove Music Online*. https://www-oxfordmusiconline
22. J Winterson, P Nickol, & T Bricheno. *Pop Music: The Text Book* (London: Edition Peters, 2013): 15.
23. Ibid.
24. Cohen, *Long steel rail*, 612.
25. Cohen, *Long steel rail*, 597.

16 Heroes and villains of the American railroads: John Henry, Casey Jones, Railroad Bill and Jesse James

THIS CHAPTER EXPLORES the lives of some of the best known heroes and villains whose exploits on the railroad were celebrated in song. Those featured have more in common than might be immediately thought, they all lived short lives in the late nineteenth-century and in each case their story is tinged with elements of both wrongdoing and heroism, whether factually correct or otherwise. The heroic Casey Jones could be, and indeed was, found guilty of dangerous driving. Strongman John Henry's feat of strength could be seen as a reckless act given that it led to his death. Both Railroad Bill and Jesse James were often hailed as Robin Hood figures, and although Railroad Bill did occasionally donate his loot to the poor, in the case of Jesse James there was no truth in this characterization at all.

John Henry

Some of the most well-known steel driving songs are about John Henry, a folk hero, strong and powerful, who could work longer than any man and who was renowned for pitting himself against the might of a new-fangled steam drill, consequently hammering himself to death. Although not all elements of this story can be verified, research has shown that as a young man John Henry may have been a steel driver on the tunnels of the Chesapeake & Ohio Railroad. In his work, *Hear My Sad Story*, the American historian Richard Polenberg (2015) recounts the true stories that inspired the song John Henry and other characters in American folk songs.

In February 1870, tracks for the Chesapeake and Ohio Railway were being laid through the Appalachians…the workers reached Big Bend Mountain, which was too high to go over and too large to go around. So a tunnel had to be cut through a mile of solid rock… When finally completed on September 12, 1872, the tunnel was 6,450 feet long, 17 feet high and 13 feet wide. The entire project took more than two and a half years and required nearly a thousand workers, some of whom were prisoners leased by the state to private contractors. Many workers were injured, others lost their lives in accidents and landslides, and still others collapsed because of the backbreaking nature of the labor. One of those who died, most likely in the fall of 1870, was a black man named John Henry.[1]

The work he was doing was hazardous. Thick smoke from the blasts and the candles used to light the tunnel became so thick that visibility was reduced to a few steps. Other dangers included falling rocks, fetid air, and the fragility of the wooden arches that supported the roof. A report in December 1870 observed that "The scene in the scantily lighted tunnel grew to resemble an inferno, men going about naked in the intense heat."[2]

In the early 1920s, the academic Louis Watson Chappell embarked on a folk-lore study of John Henry. He interviewed a few men who had worked alongside John Henry when he was working on the tunnel. Their recollections of him were very similar. All remembered him as being black, six feet tall and well-built, a great gang-leader, worker and singer. Chappell's account however, has been disputed in its details, date and location. A consensus has never been reached. This is best summarised by the American folklorist Richard Dorson when he writes of the different accounts of John Henry. 'He comes from Tennessee most often, but also from East Virginia, Louisiana, and Mobile, Alabama. His hammer weighs nine, ten, twelve, sixteen, twenty and thirty pounds; sometimes he carries a hammer in each hand.'[3] Another account that has received some credence comes from an anthropologist, MacEdward Leach, who in 1967 published an essay that sets the origins of John Henry in Jamaica rather than the USA. His account, which stems from the discovery of a song on the back of a map dated around 1894, claims that John Henry died during construction work on the Kingston - Port Antonio railroad.

Ten pound hammer it crushed me pardner
Ten pound hammer it crushed me pardner
Ten pound hammer it crushed me pardner
Somebody dying every day.

The components of the narrative usually cover John Henry's premonition as a child that steel driving would result in his death, preparation for the contest, the race against the steam hammer, his death and burial and the reaction of John Henry's wife.[4] Versions of the song are also sung to several different tunes. Here are the lyrics as sung by white American hillbilly singer Fiddlin' John Carson.

John Henry as sung by Fiddlin' John Carson

John Henry was a very small boy,
Fell on his mammy's knee;
Picked up a hammer and a little piece of steel,
"Lord, a hammer'll be the death of me,
Lord, a hammer'll be the death of me."

John Henry went upon the mountain,
Come down on the side;
The mountain so tall, John Henry was so small,
Lord, he lay down his hammer and he cried,
"Oh, Lord,"
He lay down his hammer and cried.

John Henry was on the right hand,
But that steam drill was on the left;
"Before your steam drill beats me down,
Hammer my fool self to death,
Lord, I'll hammer my fool self to death."

The captain says to John Henry,
"Believe my tunnels fallin' in."
"Captain you don't need to worry,
Just my hammer hawsing in the wind,
Just my hammer hawsing in the wind."

"Look away over yonder, captain.
You can't see like me."
He hollered out in a low, lonesome cry,
This hammer'll be the death of me,
Lord this hammer'll be the death of me."

John Henry told his captain,
"Captain you go to town
Bring John back a twelve-pound hammer,
And he'll whip your steam drill down,
And he'll whip your steam drill down."

For the man that invented that steam drill
Thought he was mighty fine,
John Henry sunk a fo'teen foot,
The steam drill only made nine,
The steam drill only made nine.

John Henry told his shaker,
"Shaker, you better pray,
For if I miss this six-foot steel,
Tomorrow'll be your buryin' day.
An' tomorrow'll be your buryin' day."

John Henry told his lovin' little woman,
"Sick and I want to go to bed;
Fix me a place to lay down, child
Got a rollin' in my head,
Got a rollin' in my head."

John Henry had a lovely little woman,
Called her Polly Ann;
John Henry got sick and he had to go home
But Polly broke steel like a man,
Polly broke steel like a man.

John Henry had another little woman,
The dress she wore was blue;
She went down the track and she never looked back,
"John Henry, I've been true to you."

In 1924 Fiddlin' John Carson (1868-1949) was the first to record 'John Henry'. This was followed by a steady stream of recordings – nearly 50 by 1948. Since then hundreds more recordings have followed, notably by blues singers Leadbelly and Mississippi John Hurt, country singer Johnny Cash, American folk singers Woody Guthrie and Pete Seeger, rock musician Bruce Springsteen, and British skiffle singer Lonnie Donegan. The appeal of John Henry stretches beyond the field of popular music. In 1940 a musical about John Henry appeared on Broadway with Paul Robeson in the lead role, however, after poor reviews it closed after only a few days. In 2009 the American composer Julia Wolfe (b. 1958) drew on different versions of the story in her piece *Steel hammer* and over 50 years earlier another American composer, Aaron Copland (1900-1990) composed *John Henry* a short work for school orchestra simulating the sounds of a train and John Henry's steel hammer.

Casey Jones

Come all you rounders[5], if you want to hear
The story told of a brave engineer
Casey Jones was the rounder's name
A high right-wheeler of mighty fame.

Train speeds had increased to at least 60 miles per hour in the 1890s and they became more difficult to control. Between 1895 and 1905, nearly 80,000 people died in railroad accidents, roughly a third of them were railroad employees.[6] One of those was Jonathan Luther Jones, nicknamed 'Casey', who died on April 30, 1900, when he tried to prevent his speeding train from crashing into another train. He saved all the passengers. His friend, an African American engine wiper named Wallace Saunders, wrote the first version of 'The ballad of Casey Jones', a tribute to the engineer. Several versions were sung over the years by countless people with hundreds of recordings being made. Casey Jones became a legend.

John Luther Jones was born on March 14, 1863, in Missouri. When his family moved to Cayce, Kentucky, the name of this town later served as his nickname, it was used to distinguish him from the many other Welsh-American Jones's. At the age of 15, he moved to Columbus, Kentucky, where he found work on the Mobile & Ohio Railroad, as a labourer then a telegrapher. From there he moved up to flagman and then brakeman, moving again, this time to

Jackson, Tennessee. Here he met Joanne Brady, known as Janie, who became his wife. In 1887 he was promoted to engineer, working for the Illinois Central Railroad. Each engineer was allowed to install his own train whistle and Casey Jones was well-known for the sound of his distinctive six-chime train whistle. In the words of the song

The switchman knew by the engine's moans
That the man at the throttle was Casey Jones.

In January 1900 Jones took over the controls of the *Cannonball Express*, a fast passenger train running from Chicago to New Orleans. On Sunday April 29, 1900, Casey Jones and his fireman Sim Webb were due to take the *Cannonball Express* on its 188-mile run south to Canton, but the train arrived 95 minutes behind schedule and they had to set off early the following morning. Jones had a reputation as a 'fast roller' and they were determined to make up for lost time, and indeed they did regardless of the darkness, the fog and the rain.[7] They were almost on schedule and close to their final destination, but because of unusual circumstances the caboose[8] of a freight train was in the way on the mainline. As Jones caught sight of the lights of the caboose, he told Webb to jump as he tried to apply the brakes. The following is how the fatal crash was reported on May 1 1900 on the front page of the *Memphis Commercial Appeal*.

DEAD UNDER HIS CAR
THE SAD END OF THE ENGINEER CASEY JONES

ILLINOIS CENTRAL WRECK

Southbound Passenger Train No. 1 Crashes into the Rear of a Freight – Details of the accident.

Jackson, Miss. April 30 – (Special) - A disastrous collision occurred about 4 o'clock this morning on the Illinois Central Railroad at Vaughan, a station eleven miles north of Canton. The Engineer, Casey Jones, was instantly killed and Express Messenger Miller was hurt internally, but not seriously.

The south-bound passenger train was running under a full head of steam when it crashed into the rear end of a caboose and three freight cars which were standing on the main track… The caboose and two of the cars were smashed to pieces, the engine left the rails and plowed into an embankment, where it overturned and was completely wrecked, the baggage and mail coaches also being thrown from the track and badly damaged. The engineer was killed outright by the concussion. His body was found lying under the cab with his skull crushed and the right arm pulled from the socket. The fireman jumped just in time to save his life. The express messenger was thrown against the side of the car, but his condition is not considered dangerous.

The other employees and all of the passengers were more or less jolted by the shock, some of them receiving bruises and slight wounds, none of which, however, were serious.

Every effort was made to stop the speeding train, but without success. Two flagmen were sent down the track with danger signals and torpedoes were placed on the rails as a warning, but the engineer did not seem to take any notice of the signals nor to realise the situation until a short distance of the caboose, when he made a violent attempt to put on the airbrakes, but the distance was too short to avoid a crash. The freight boxes were loaded with bundled hay and the scattered coal from the engine soon set fire to the debris, and it was feared at one time that the whole mass of wreckage would be destroyed, but the fire was eventually extinguished without doing very great damage…[9]

This is essentially what happened.[10] Although rather than dying of a crushed skull, Sim Webb later recalled that 'they found Casey with one hand clutching the throttle and the other the air-brake control' and that Jones died because either a metal bolt, or a piece of splintered wood, pierced his throat. Although the *Memphis Commercial Appeal* report concluded with the words that 'Engineer Jones …was highly esteemed as one of the road's safest and most capable engineers',[11] this was not the view of the officials of the Illinois Central Railroad company who had found his safety record wanting in previous years and had suspended him no less than nine times in violation of rules. The company investigation following the fatal crash came to the unequivocal conclusion that 'Engineer Jones was solely responsible for the accident as consequence of not having properly responded to flag signals'.[12] His body was sent back to Jackson, Tennessee and a funeral mass was held the next day at the church where he had married Janie Brady 14 years earlier. Railroad men came from miles around to attend the funeral and pay their last respects. In the aftermath, however, Jones became the legend who had unselfishly given his life so that everyone on his train would live.

When Jones's friend Wallace Saunders composed the earliest ballad about Casey Jones, he used a popular tune, 'Jimmie Jones' but he did not copyright the lyrics. One verse tells of Jones' resolve to reach his destination on time.

Fireman say, "You running (too) fast,
You ran the last three lights we passed"
Casey say, "We'll make it through,
She's steamin' better than I ever knew"

In 1901 two vaudeville performers, T. Lawrence Seibert and Eddie Newton, were the first to copyright the song when they published their version which they called 'Casey Jones, the Brave Engineer'. One of the earliest recordings of this version was by Billy Murray in 1911. This included a verse which suggested that Mrs. Jones had found someone to replace Casey.

Miss Casey Jones setting upon the bedside
When she received the message that her Casey had died
Says, go to bed chillun' and ahush your crying
For you've got another poppy on the Salt Lake Line.

Not surprisingly she was offended and spent many years disclaiming the idea that she was having an affair.[13] Murray's recording was so popular that within 20 years of Jones's death, many versions of the song had been recorded, millions were sold, and the sheet music was widely available. Making the most of the song's popularity, a parody, 'Casey Jones—the Union Scab', appeared in 1911. The words were written by Joe Hill, an organizer of the radical union Industrial Workers of the World at the start of a walkout of 40,000 railway workers. The lyrics told of a railroad worker who refused to support a strike, then died and went to heaven to be told by St. Peter that he could get a job as a scab. It was largely forgotten until Pete Seeger recorded it in 1941 on an album called *Talking Union*. Carl Sandburg included the song in his book of American folk songs, and described it as the 'greatest ballad ever written.'[14] The first blues singer to record it was Mississippi John Hurt but the recording was never issued. Instead it was the black singer Walter "Furry" Lewis's who introduced the song to blues audiences when he recorded it in 1928 with the title 'Kassie Jones', deliberately misspelled to avoid copyright infringement. The lyrics include the well-known chorus

On the road again,
I'm a natural born eas'man, on the road again.

There are now hundreds of recordings of the song and many variations of it including another take from bluesman Jesse James, 'Southern Casey Jones' (1936) and a much later Grateful Dead song 'Casey Jones' (1970) written by Jerry Garcia (see pages 269–70).[15]

Railroad Bill

Railroad Bill ought to be killed
Never worked and he never will
Now I'm gonna ride my Railroad Bill

Morris Slater (the real name of Railroad Bill) was born a slave in North Carolina. In the 1890s he became notorious as a train robber, mostly in southern Alabama on the Louisville & Nashville Railroad (L&N), a thriving company with more than 2,614 miles of track, hundreds of locomotives and passenger cars, and thousands of freight cars.[16] Slater's first crime took place in 1896 when he shot at the deputy sheriff Allen W Brewton in Bluff Springs, Florida. Brewton had told Slater that he needed to buy a licence for his Winchester rifle or to give it up. Brewton fired at him and missed, Slater fired back and then disappeared into the swamp to become a wanted man. He remained at large for more than three years, moving around the area, riding boxcars, robbing trains and freight stations, and shooting it out with the law. Sometimes he kept the stolen goods and sometimes he gave them away or sold them off cheap to the local poor. To many blacks in Alabama he became a folk hero, but at the same time he became the object of a manhunt. One of those out for Slater's blood was Sheriff Edward McMillan. In 1895 he received a tip off and located Slater in a house near Bluff Springs. The shooting that followed resulted in McMillan's death and again Slater slipped out of sight. Some versions of the song include the following verses:

Railroad Bill made a mighty dash
Shot McMillan by a lightning flash
Talkin' bout that Negro, Railroad Bill.

Railroad Bill, goin' down the hill
Lightin' cigars with a five-dollar bill
And its ride, ride, ride.

A reward of $1,250 was offered for information leading to Slater's arrest, $350 of it was pledged by the L&N railroad, which also threw in a free railroad pass for life. Consequently the size of the search parties increased, 'harassing, arresting, or beating blacks, blameless though they may have been, who were suspected of shielding Slater'.[17] Slater remained at large committing further train robberies, often breaking into a railroad car, throwing the valuables onto the track with the train still in motion and then returning to collect the loot. Some years later a report was written by J B Harlan who had been working for the L&N police and had been assigned to capture Slater. Naturally he was entirely without sympathy for Slater, but he made it clear that the view of the local black population was different.

> it was generally believed and talked by the negroes in that section of the country that "Railroad Bill" was superior being, possessed of super-human powers and, having been shot at so many times by different raiding parties, that he could not be killed with an ordinary leaden bullet; some of them claiming that the only bullet that would kill him was a solid silver missile.[18]

Eventually the law caught up with Slater in March 1896 at a general store in Atmore, Alabama, where he sat armed with a Winchester rifle and two loaded pistols and eating stolen cheese and crackers. Accounts of what exactly happened next differ, but the shootout resulted in Slater's death.

Railroad Bill lyin' on de grocery floor
Got shot two times an' two times more
No more lookin' fer Railroad Bill.

His body was embalmed and then exhibited far and wide to large crowds who paid an admission fee to see it. Not long after his death, songs started to be written about Railroad Bill. Verses were collected as early 1909 and were published a couple of years later by E. C. Perrow and by Howard W. Odum. Some versions have few direct references to Slater's train robbing history, rather they refer to him as an idle womaniser. The version included in Carl Sandburg's *American Songbag*, for example, includes the following verses.[19]

Railroad Bill, Railroad Bill,
He never work and never will;
Well, it's bad Railroad Bill.

Railroad Bill had no wife,
Always looking for somebody's wife;
Then it's ride, ride, ride.

Railroad Bill, might bad man
Shot the lantern out the brakeman's hand
Bad Railroad Bill.

The first recordings appeared in 1924 by Roba and Bob Stanley, as well as Gid Tanner, Riley Puckett, and the Skillet Lickers. 'Railroad Bill' songs became more widespread during the urban folk revival and its popularity continued with recordings by Joan Baez, Lonnie Donegan, Bob Dylan, Ramblin' Jack Elliott, and Sonny Terry and Brownie McGhee.

Jesse James

It was on a Wednesday night, the moon was shining bright
They robbed the Glendale Train.
And people they did say, for many miles away
'Twas those outlaws Frank and Jesse James.

Chorus
Jesse had a wife to mourn all her life,
The children they are brave,
'Twas a dirty little coward shot Mister Howard
And laid Jesse James in his grave.

It was Robert Ford, the dirty little coward.
I wonder how he does feels,
For he ate of Jesse's bread and he slept in Jesse's bed,
Then laid Jesse James in his grave.

It was with his brother Frank he robbed the Gallatin Bank,
And he carried the money from the town.
It was in this very place they had a little race,
For they shot Captain Sheets to the ground.

They went to the crossing not very far from there,
And there they did the same.
And the agent on his knees, he delivered up the keys
To the outlaws Frank and Jesse James.

It was on a Saturday night, Jesse was at home
Talking to his family brave,
When the thief and the coward, little Robert Ford,
Laid Jesse James in his grave.

How people held their breath when they heard of Jesse's death,
And wondered how he ever came to die.
'Twas one of his gang, dirty Robert Ford
That shot Jesse James on the sly.[20]

Jesse James was a murderer and an outlaw, arguably the most famous outlaw in American history. He was born in 1847 in Kearney, Missouri, the son of a Baptist preacher. His older brother Frank fought on the side of the South in the American Civil War and was a member of the notorious Quantrill's Raiders who in Lawrence, Kansas, in August 1863 engaged in an orgy of looting and murder. The following year Jesse James, aged 16, joined his brother in the gang. When the Quantrill Raiders broke up in 1865 the brothers formed the James-Younger gang with other ex-guerrillas and turned to robbing banks – seven in the next three years, the first being the Clay County Savings Association in Missouri. As they made their getaway, one of the robbers shot and killed a bystander. In another bank raid in Gallatin, Missouri, James mistakenly thought that one of the bank clerks was his nemesis Lieutenant Colonel Samuel P. Cox and shot him through the heart.

After laying low for over a year they committed a further 12 robberies in just over three years. In 1873 they turned their attention to railroads. The Chicago, Rock Island and Pacific Railroad was their first target. The gang derailed the train and broke into the safe. The engineer died in the crash and some passengers were injured. In July 1876 the James gang stopped the Missouri Pacific Railroad train and made off with thousands of dollars. Their next target was the Chicago & Alton Railroad at Glendale, Missouri. Between 1880 and 1881 there were four more hold-ups but, unknown to James, the gang had been infiltrated by two brothers, Robert and Charles Ford, who had been offered a pardon and a reward for the murder of Jesse James by the governor of Missouri, Thomas T. Crittenden. The Ford brothers were related to the James family and had been invited to stay with them. On the morning of April 3, 1882 Jesse James climbed on a chair to dust a picture and was shot in the back by Robert Ford, he died instantly. He was buried at his mother's home, and his grave became a shrine with countless viewers paying to see it.

James had already been built into a hero during his lifetime, a latter day Robin Hood, but there is no evidence to support any such good deeds. In reality he was a train robber who had little or no sympathy for his victims and killed several people along the way. James himself helped to create this sympathetic image, writing to newspapers attempting to justify his motives and covering a range of excuses: sometimes portraying himself as robbing the rich to give to the poor; sometimes saying that he was acting in self-defence;[21] and even claiming that he had to rob in order to pay his heavy taxes.[22] So why was the general population so sympathetic to James? Norm Cohen argues that there was a general antipathy towards the railroads going back to the 1870s: farmers had felt cheated in their land deals; some had lost their savings in railroad stocks and bonds; many felt that the freight rates were unfair and that they had to pay over the odds to transport their products to the market. There was a similar antipathy to the banks: farmers had had to borrow large sums to buy new machinery and when agricultural prices fell some were unable to pay their mortgages. Consequently there was little sympathy for what was perceived as the wealthy victims. All this helped to make the holdups a popular crime.[23]

There has been much debate about the authorship of the song. For a full discussion see Norm Cohen's *Long steel rail: the railroad in American folksong*.[24] The earliest version of a song about Jesse James was published in 1887 in a small book of songs, and a few years later the verses appeared on a New York City broadside. In 1910 the song was published in John A. Lomax's *Cowboy Songs and Other Frontier Ballads*. The Lomax version portrays Jesse James as a friend to the poor, who 'never would see a man suffer pain.'

The first recording of 'Jesse James' was made by Bentley Ball in 1919. Another early recording was made by Riley Puckett in 1925, included the sympathetic line 'Dear old Jesse, poor old Jesse' and a sign of repentance, 'He fell down upon his knees and he handed up the keys / To the bank that he had robbed the day before.' Harry McClintock's 1928 recording was again on the side of the outlaw and included the lines 'he took from the rich and he gave it to the poor'. In 1939 Woody Guthrie followed the same line claiming that the brothers 'never was outlaws at heart'; 'the railroad bullies come to chase them off their land'. Many recordings have followed notably by Pete Seeger, The Pogues, Van Morrison and Bruce Springsteen.

Endnotes

1. Richard Polenberg. *Hear My Sad Story: The True Tales That Inspired "Stagolee," "John Henry," and Other Traditional American Folk Songs*. (New York: Cornell University Press, 2015): 150.
2. Cited in Polenberg, *Hear My Sad Story*, 149.
3. Richard M Dorson. 'The Career of "John Henry"'. *Western Folklore*, July, 1965, Vol. 24, No. 3. (Long Beach California: Western States Folklore Society): 162.
4. Norm Cohen. *Long steel rail: the railroad in American folksong*. (Urbana: University of Illinois Press, 2000): 72.
5. 'Rounders' is an American term referring to itinerant railroad men.
6. Polenberg, *Hear My Sad Story*, 163.
7. Cohen, *Long steel rail*, 136-7.
8. A caboose is a manned North American railroad car coupled at the end of a freight train.
9. May 1 1900, *Memphis Commercial Appeal*.
10. Cohen, *Long steel rail*, 137.
11. *Memphis Commercial Appeal*, May 1, 1900.
12. Polenberg, *Hear My Sad Story*, 166.
13. Cohen, *Long steel rail*, 144.
14. Carl Sandburg. *The American Songbag*. (New York: Harcourt Brace Jovanovich, 1927).
15. For a full discussion of the many versions and recordings of the song see Norm Cohen pp 132-157.
16. Polenberg, *Hear My Sad Story*, 130-131.
17. Polenberg, *Hear My Sad Story*, 133.
18. Cited in Cohen, *Long steel rail*, 124.
19. Sandburg, *The American Songbag*, 384.
20. Sandburg, *The American Songbag*, 420.
21. Cohen, *Long steel rail*, 102.
22. Polenberg, *Hear My Sad Story*, 115.
23. Cohen, *Long steel rail*, 102.
24. Cohen, *Long steel rail*, 103- 107.

17 Trains, lines and wrecks on the early American railroads

AMERICAN TRAINS DIFFERED from their European counterparts in several ways; the trains were bigger and heavier, the locomotives stronger and larger and, because there were few tunnels and bridges, the trains could often be taller. Paul Oliver (1990) writes of the way that trains were 'Impressive in their speed and immense proportions, chilling the spine with their shrieking whistles in the night, thrilling the blood with the roar of their engines as they passed…'[1] Not surprisingly, as we have seen in a previous chapter, some trains, such as the *Dixie Flyer* and the *M&O*, took on a particular significance and were named in song. This chapter looks at further trains that were celebrated in song, as well as one of the railroad lines whose fame became widespread in the 1950s through a promotional song of that name written in the 1920s. Not all of the songs where trains are named in their titles were written in celebration, rather they were written in commemoration of the many who died in train wrecks in the early decades of the railroad, probably the most famous being 'The wreck of the Old 97'.

The Wabash Cannonball

The term 'cannonball' was used to mean a fast train. There were several Wabash Cannon Ball passenger trains as early as the 1880s on the Wabash Railroad; one between St. Louis and Omaha and was renowned as the fastest train of its day taking its 415 mile trip at an average of 27 miles per hour.[2] The song tells of a train on an imaginary route across America. It is an impossible route as it goes off in different directions. In the *Hobo's Hornbook*, George Milburn writes

"'The Wabash Cannonball' is for the hobo what the spectral 'Flying Dutchman' is for the sailor. It is a mythical train that runs everywhere, and the ballad about it consists largely of stanzas enumerating its stops'.[3] According to an anecdote recorded by Alan Lomax (1975) the train went so fast that after it had been drawn to a halt it was still travelling at 65 miles an hour, it had 700 cars and the conductor punched holes in the tickets with a 45 calibre automatic.[4]

The Wabash Cannonball

From the Rocky bound Atlantic to the wild Pacific shore,
From the coast of Maryland to the ice-bound Labrador,
There's a train of splendour and it's quite well known to all,
The modern 'ccommodation called the Wabash Cannonball.

Great cities of importance that we reach upon our way,
Chicago and St Louis, Rock Island so they say,
Springfield and Decatur, Peoria and them all,
We reach them by no other than the Wabash Cannonball.
You can hear the merry jingle and the rumble and the roar
As she dashes through the woodland, comes creeping long the shore,
We hear the engine's whistle and the merry hoboes call,
As they ride the rods and brake-beams on the Wabash Cannonball.

'The Wabash Cannonball' evolved from the folk song 'The Great Rock Island Route', first published in 1882. Although several recordings had been made earlier, the two most well-known were released in 1937 and 1947 both featuring the country singer Roy Acuff (1903 – 1992. Acuff appeared on the earlier version making imitation train whistle sounds, a trick he had learnt as a child. He recorded more than a dozen railroad songs and in 1967 released a whole album of them which includes 'Night train to Memphis', 'Sunshine Special', 'Wreck of the Old 97' and 'Freight train blues'.

Midnight Special

The song *Midnight Special* originated among prisoners in the American South. It refers to a passenger train and its ever-loving light (sometimes 'ever-living light'). There are many different versions, some of them including lines from other prison songs, but most include the following lines

O let the Midnight Special
Shine a light on me,
Let the Midnight Special shine a light on me
Shine a evah lovin' light on me!

According to Stephen Wade (2012), the 'Midnight Special' travelled to Parchman Prison (Mississippi State Prison) on visitor's day which was scheduled in any months that had five Sabbaths. The train departed from Jackson, Mississippi shortly after midnight and brought the friends and families of the prisoners, arriving at dawn and returning at dusk. In 1932, Mississippi Governor Mike Conner had established parole hearings which he referred to as the 'mercy court'. Visits on the Midnight Special presented an opportunity for women to ask for their men to be pardoned. Some believed that those bathed in the headlight of the train would get their pardon or that the inmate who first saw the train's light would be freed. The song 'Midnight Special' was recorded at Parchman in 1937 but it was also found in other penitentiaries.[5] According to Alan Lomax, the Midnight

Rail road workers standing on a train, 1890-1895
Buncombe County Special Collections, Pack Memorial Public Library, Asheville, North Carolina, USA

Special was the train officially named the Southern Pacific Golden Gate Limited which, as it travelled from Houston, shone its light on Sugar Land Prison in Texas. The inmates believed that if the light shone on them it was a sign that they would be freed. Leadbelly popularized the song upon his release from Sugar Land.[6]

Carl Sandburg's 1927 collection *The American Songbag* includes two versions of 'Midnight Special'.[7] Both refer to the pardon and one of these has the following verse

Yonder come Miss Rosie;
Oh, how do you know?
By th' umbrella on her shoulder
An' the dress that she woah!
Piece of paper in her hand
Says, "Look here, Mr. Jailer,
I wants my life-time man."

The song was first commercially recorded in 1926 as 'Pistol Pete's Midnight Special' by Dave "Pistol Pete" Cutrell with McGinty's Oklahoma Cow Boy Band. In 1934 Leadbelly recorded a version of the song at Louisiana State Prison in Angola for John and Alan Lomax. Leadbelly went on to record at least five versions of the song, Since then many other blues artists have made recordings including Big Bill Broonzy and Sonny Terry and Brownie McGhee, but it has also been covered in various other styles with bands such as ABBA, Van Morrison, Johnny Rivers and, probably the most well-known, Creedence Clearwater Revival.

Another train that was celebrated in song is the 'Flyin' Crow'. The blues singer Washboard Sam recorded the song in the 1930s using the washboard to convey the rhythm of the train flying along.

Flyin' Crow

Flyin' Crow (as sung by Washboard Sam)

Flyin' Crow, leave Port Arthur
Come to Shreveport to change her crew (twice)
She will take water at Texarkana,
Yes, boys then keep on through

That Flyin' Crow, whistle
Sounds so lonesome and sad (twice)
Lord, it broke my heart
And took the last woman I had

Now, she's gone, she is gone
With a red and green light behind (twice)
Now she's gone, she is gone
The red is for trouble
And the green is for my ramblin' mind.[8]

Freight train

Many of the songs covered in the chapters about American folk songs and blues have complicated histories: they exist in different versions; some are set to more than one melody; and it is often difficult to trace their provenance. There is little ambiguity with the two songs 'Freight train' and 'Freight train blues' however. We know that 'Freight train' was composed by Elizabeth Cotten and her original music is adhered to.

Freight train, freight train, run so fast
Freight train, freight train, run so fast
Please don't tell what train I'm on
They won't know what route I'm going.

When I'm dead and in my grave
No more good times here I crave
Place the stones at my head and feet
And tell 'em all that I've gone to sleep.

When I die, Lord bury me deep
Way down on old Chestnut Street
So I can hear old Number Nine
As she comes rolling by.

Elizabeth Cotten (1895-1987) was born in Chapel Hill, North Carolina in 1893. She was a musically gifted child and taught herself to play, borrowing her brother's banjo and guitar in secret. When she went to work at the age of 12 she saved up money to buy her own instrument and started to write her own songs including 'Freight train'.

> We used to watch the freight train. We knew the fireman and the brakeman… and the conductor…They'd let us ride in the engine…put us in one of the coaches while they were backing up and changing…that was how I got my first train ride. We used to walk the trestle and put our ear to the track and listen for the train to come. My brother, he'd wait for the train to get real close and then he'd hang down from one of those ties and swing back up after the train had passed over him.

At the age of 15 she married and had a daughter. After undergoing a religious conversion, she became more interested in the church and soon after she more or less gave up playing the guitar for about 25 years. She moved to Washington, DC, in 1945 and it was there that she met the Seeger family of folk musicians. She was working in a department store when she came across a lost little girl – Peggy Seeger – and returned her to her parents the composer Ruth Crawford Seeger and the musicologist Charles Seeger. She became friends with the family and started to work for them. One day she picked up a guitar at the Seeger's home and started to play in her inimitable finger-picking style: she was left-handed and, rather than re-stringing the instrument, she played the guitar upside down playing the treble strings with a thumb and the bass strings with her fingers. Her playing was encouraged by the Seeger family: in 1957, Mike Seeger produced her first album, *Folksongs and Instrumentals with Guitar* and in 1959 she gave her first public concert, followed by performances on the folk circuit. This led to a concert career where she would talk about her life and sing with guitar or banjo accompaniment. In the early 1960s, she went on to play concerts with some well-known blues singers including Mississippi John Hurt, John Lee Hooker and Muddy Waters. Through the success of these concerts she was able to buy a house in Syracuse, New York. Cotten recorded 'Freight train' in 1958 but it was popularised by Peggy Seeger and soon became a standard on the folk revival circuit.

Freight train blues

'Freight train blues' was composed by John Lair in about 1934 and was popularised by country singer Roy Acuff's recording. The song is very familiar to many through Bob Dylan's 1962 debut album *Bob Dylan*. In a radio interview Lair recalled the lonesome sound of the train whistle in the night, disturbing the quiet countryside of the Kentucky mountains where he was brought up. Dylan copyrighted his arrangement in 1978 but the Acuff and Dylan versions have much in common including the train whistle imitation on the word 'blues' at the end of the chorus where there is a long descending slide.

I was born in Dixie in a boomer's shack
Just a little old shanty by a railroad track
The hummin' of the drivers was my lullaby
And a freight train whistle taught me how to cry.

Chorus
I've got the freight train blues, Lordy, Lordy, Lordy
Got 'em in the bottom of my ramblin' shoes
And when that whistle blows, I've gotta go
Oh Lordy, guess I'm never gonna lose
The mean old freight train blues.
Now my pappy was a fireman and my mammy dear
Was the only daughter of an engineer
My sister married a brakeman and it ain't no joke
Now it's a shame the way she keeps a good man broke.

The Rock Island Line

The railroad that became famous as the Rock Island Line ran from Chicago to Rock Island in Mississippi and was completed in 1854.[9] There were eight thousand miles of track across fourteen Midwestern states and the company employed over 41,000 workers. The Chicago, Rock Island and Pacific Railroad encouraged its workers to take part in activities such as singing in choirs and writing songs to help promote the line and in 1920 they created a network of what they referred to as booster clubs. The original 'Rock Island Line' was written by Clarence Wilson, an engine cleaner in the central freight yard at Little Rock in Arkansas, who recorded it with his gospel quartet of fellow workers, the Rock Island Colored Booster Quartet in 1929. The lyrics to Wilson's song,

'Buy your ticket over Rock Island lines', included verses describing people who worked at the yard.

Engineer Kugler is a good one, too.
He blows his whistle 'til it makes you blue,
He runs a freight train on passenger time,
Be sure and buy your ticket over the Rock Island Line.

This referred to George J. Kugler, known by many as the Musical Engineer because he had created a steam whistle attachment which meant that he could control the pitch of the train whistle and play whole melodies. On leaving the station he often played 'Goodbye my lover goodbye' for his wife and then signalled his return with his homecoming tune 'Polly put the kettle on.'[10] Most of Wilson's lyrics are quite different to the versions that later evolved, but the chorus has the following familiar promotional lyrics

The Rock Island Line is a mighty good road
The Rock Island Line is the road to ride
The Rock Island Line is a mighty good road
If you want to ride you gotta ride it like you find it
Get your ticket at the station for the Rock Island Line.

The first recording of 'Rock Island line' was made by folklorist John Lomax in 1934 at Cummins State Prison in Arkansas. Stephen Wade (2012) in his study of field recordings writes that it was sung by a 21-year-old convict, Kelly Pace, who led the song with a group of his fellow inmates singing in three and four-part harmony, using call-and-response and train whistle imitations between the choruses. Three years later Leadbelly made his first recording of the song; this version was in the manner of a work song depicting wood chopping. He then went on to record at least five solo versions of the song. These percussive songs with 12-string guitar accompaniment depict a freight train express bound for New Orleans with a cargo of livestock and pig iron, a contest between an engineer and a depot agent, with train whistles and calls. Leadbelly creates a rhythmic showpiece, far removed from the gospel quartet that the Arkansas prisoners performed. The song includes a spoken introduction telling of a train operator smuggling pig iron through a toll gate and claiming that he only had livestock on board. Here are the lyrics of Leadbelly's version.[11]

Rock Island Line

Chorus
On the Rock Island Line, it's a mighty good road
On the Rock Island Line, there's a road to ride
On the Rock Island Line, it's a mighty good road
If you want to ride, you gotta ride it like you find it
Get your ticket at the station on the Rock Island Line

Jesus died to save our sin
Glory to God, we're goin' to meet Him again

Chorus

I may be right and I may be wrong
No, you are goin' to miss me when I'm gone

Chorus

Yes, the A B C double X Y Z
Cats in the cupboard but they can't see me

'Rock Island Line' became a huge hit for Lonnie Donegan and his Skiffle Group in 1955. Donegan's band was made up of acoustic guitar, double bass (Chris Barber) and washboard (Beryl Bryden). This DIY line-up signalled the start of a skiffle craze and went on to inspire many future rock musicians including the Beatles and the Rolling Stones. The song 'John Henry' featured on the B side of Donegan's recording. Donegan took out the copyright in the song. *The Penguin Book of American Folk Songs* (1964) edited by John Lomax's son Alan Lomax includes the following footnote to Rock Island Line:

> John A. Lomax recorded this song at the Cummins State Prison farm, Gould, Arkansas, in 1934 from its convict composer, Kelly Pace. The Negro singer, Lead Belly, heard it, rearranged it in his own style, and made commercial phonograph recordings of it in the 1940s. One of these recordings was studied and imitated phrase by phrase, by a young English singer of American folk songs [referring to Lonnie Donegan], who subsequently recorded it for an English company. The record sold in the hundreds of thousands in the U.S. and England, and this Arkansas Negro convict song, as adapted by Lead Belly, was published as a personal copyright, words and music, by someone whose contact with the Rock Island Line was entirely through the grooves of a phonograph record.[12]

There have been many further recordings, most notably by Johnny Cash, Woody Guthrie, Bobby Darin, Little Richard, and Pete Seeger, but no one has repeated Lonnie Donegan's gold record success.

Railroad accidents

In 1853 there were eleven major train collisions and derailments in the USA resulting in 121 deaths. Cohen (2000) reports that the earliest surviving American broadside ballad tells of a rail disaster on the North Pennsylvania Railroad in 1856 where an excursion train carrying 1000 Sunday school children was in a collision where 66 died. No train wreck ballads survive from the 1860s and 1870s. By 1900 the railroads employed over a million men and women and for those who worked on the lines, accidents at work were the single most frequent cause of death. Railroad accidents also killed hundreds of passengers each year, and those who lost their lives walking on tracks or hopping freight trains mounted to thousands. In all nearly 12,000 people had died on the rails by 1907. Folklore thrives on danger and disasters and there were many songs describing train disasters. The crew was usually the focus of the lament and a typical song was a tribute to the crewmen, often named, who had lost their life in the course of duty.

10 songs about train wrecks

SONG TITLE	THE ACCIDENT	DATE
The wreck on the C & O	The *Fast Flying Virginian* (FFV) struck a rock on the road near Clifton's Ford, Virginia. The engineer was killed but the fireman leapt to safety.	1890
The wreck of the Old 97	The fast mail train Number 97, flew off a bridge near Danville, Virginia, the locomotive and five cars fell into a ravine 75 feet below. Nine crew members were killed.	1903
The New Market wreck	The conductor misread an order leading to a head-on collision with either 56 or 62 killed including both engineers.	1904

The wreck on the C & O Number Five	The Number Five, a luxury passenger train known as the *Sportsman*, hit a broken line and was thrown into a bank near Dickson, West Virginia. The engineer was killed but the fireman leapt to safety.	1920
The freight wreck at Altoona/ The wreck of the 1262	Brake failure caused the train to run out of control near Altoona, Pennsylvania. It sped downhill at about 60 miles an hour and derailed. The engineman and the fireman were both killed.	1925
The wreck of the 1256	The engine ran into a landslide and overturned into the James River near Clifton Forge, Virginia. The engineer died but the crewman, whose cab was submerged in water, was rescued by a group of hoboes.	1925
The wreck of the Royal Palm	19 people were killed in a collision between the *Royal Palm* and the *Ponce de Leon* near Rockmart, Virginia. A coach flew up into the air and fell on to the diner of the *Ponce de Leon*.	1926
The wreck of the Virginian Number Three	Two Virginia Railway trains, a passenger train and a freight train, collided as a result of the passenger crew's failure to meet an order. The engineer and the fireman from the passenger train were scalded to death.	1927
The wreck of Number Nine	Fictional account where, following a fatal collision, the dying engineer asks that his house should be left to his sweetheart.	Published in 1927
The wreck of Number Four	Passenger train Number Four was derailed at Torrent, Kentucky on the Lexington and Eastern line. After rounding a sharp curve it overturned killing the engineer.	1928

Mark Aldrich (2006) in *Death Rode the Rails: American Railroad Accidents and Safety, 1828-1965* writes that derailments rose steadily after 1897, and then jumped from 3,633 in 1902 to 7,432 in 1907. Collisions too shot up from 5,042 in 1902 to 8,026 by 1907. 1920 saw the largest number of derailments, 22,477, mostly involving freight trains. From then on, as a result of the development of safer technology and major investments in roadbed, derailments fell steadily—to 9,871 in 1929. In 1903 the derailment of the Southern Fast Mail was immortalized in the song 'The Wreck of the Old 97'.

The wreck of the Old 97

Well they gave him his orders at Monroe, Virginia,
Said: "Steve, you're way behind time,
"This is not 38, this is Ol' 97,
"Put her into Spencer on time."

Then he turned around and said to his black, greasy fireman,
"Shovel on a little more coal.
"And when we cross that White Oak mountain,
"Watch Ol' '97 roll."

'Cos he was going down a grade making 90 miles an hour,
The whistle broke into a scream.
He was found in the wreck with his hand on the throttle,
Scalded to death by the steam.

And then a telegram come from Washington station,
This is how it read:
"Oh that brave engineer that run Ol' 97,
"Is lyin in old Danville dead."

Oh, now all you ladies you'd better take a warning,
From this time on and learn.
Never speak hard words to your true-lovin' husband.
He may leave you and never return.

Among the casualties was the engineer, Joseph Andrew "Steve" Broady. On the morning of the accident Broady was driving Fast Mail Train No. 97 out of Monroe, Virginia, heading toward Spencer, North Carolina. He was an hour behind schedule by the time he departed and tried to make up time. Air brakes became mandatory on US trains in 1893 and No. 97 was equipped with a Westinghouse air brake system. However, applying the brakes too often without allowing air to build back up in the reservoirs could result in brake failure. Broady's engine leapt off the track and flew off the bridge into the air, dragging the other wooden cars behind it. Eleven people were killed, along with engineer Broady. These included the fireman, the conductor, three mail clerks and the flagman. A further seven were seriously injured. A few years after the crash the families of those involved in the accident took on the officials of the Southern Railroad in pursuit of compensation. Some families were given monetary awards but when Broady's relatives claimed that the railroad trestle was known to be unsafe, the Southern Railroad attested that the blame for the disaster must rest with Broady himself

This is how the accident was reported in *The Washington Post* the next day, September 28, 1903.

NINE FELL TO DEATH. Mail Train Plunged from Seventy-foot Trestle.

While running thirty or thirty-five miles an hour, train No. 97, on the Virginia Midland branch of the Southern Railroad, jumped from a trestle seventy-five feet high, half a mile north of Danville, Va., this afternoon, and was almost demolished. The wrecked train was exclusively for mail and freight, consisting of four postal cars and one express car, and was southbound. Of the sixteen persons on the train, nine were killed and seven injured.

The trestle where the accident occurred is 500 feet long and is located on a sharp curve. The engine had gone only about fifty feet on the trestle when it sprang from the track, carrying with it the four mail cars and an express car. The trestle, a wooden structure, also gave way for a space of fifty feet. At the foot of the trestle is a shallow stream with a rocky bottom. Striking this, the engine and cars were reduced to a mass of twisted iron and steel and pieces of splintered wood. All the men were killed instantly, it is thought, and all were greatly mutilated. The skin and hair of the engineer and firemen were torn off by the impact of the steam engine.

Several thousand people were soon at the scene of the wreck. No one on any of the cars had made an effort to jump, and the bodies of all those killed were found in the wreckage of the different cars to which they belonged. Ladies who drove out to the wreck from Danville fainted at the sight of the bodies. It seemed miraculous that any one should have escaped, for each car falling with the engine bounded from it and completely collapsed after striking the rocks at the foot of the trestle. All the express in the express car was practically destroyed except six crates full of canary birds. None of the birds was hurt, though the crates were in the thickest of the debris…Unofficial opinion is that the cause of the wreck was the high speed of the train on the sharp curve. Railroad men believe a flange on the engine wheel broke. Train No. 97 was running about an hour behind time…Broady, the dead engineer, was about fifty-five years of age, and had been with the Southern Railroad about twenty years, his service a large part of the time being on the division on which the accident occurred.[13]

Photographs taken from above the scene ran in newspapers across the country. Capitalising on the morbid fascination of the public with tragedies and disasters, the local newspaper, the *Danville Bee* reported

The bodies of the dead men were broken almost entirely in pieces and horribly mangled, particularly about the head and face. The impact of the steam against the bodies of the engineer and fireman caused the skin and hair to fall away from their bodies.... All the cars except one are battered into kindling wood.[14]

In 1923, Henry Whitter, a cotton mill worker from Virginia was the first to record the railroad song with the title 'The Wreck of the Old Southern 97'. A year later, the light opera singer turned hillbilly singer, Vernon Dalhart, made what became the most popular early recording selling millions over the next few years[15].

The melody comes from 'The ship that never returned' written in 1865 by Henry Clay Work, the composer of 'Grandfather's clock', but there are many different versions of the lyrics. Their origins have long been debated. In 1924 the song was copyrighted in Whitter's name prompting a long series of legal battles over who had written the song and who was entitled to royalties from its sale. In 1927 Danville resident David Graves George, a telegraph operator who had witnessed the accident scene, wrote to the *Richmond News Leader* claiming that he had written the song and in 1933 he filed a suit against the record company Victor, claiming authorship. The court ruled that George was the song's original author. In 1934 The Third Circuit Court of Appeals reversed the lower court's ruling but later that year the U.S. Supreme Court ruled in favour of George. In 1938 he was awarded over $65,000, but in 1939, the decision was yet again reversed and the legal wrangling finally came to an end.[16] There have been many recordings since.

Endnotes

1. Oliver, *Blues Fell*, 65.
2. Norm Cohen. *Long steel rail: the railroad in American folksong.* (Urbana: University of Illinois Press, 2000).
3. George Milburn. *The Hobo's Hornboook.* (New York: Ives Washburn, 1930): 189.
4. Lomax. *The Folk Songs of North America.* (New York: Doubleday): 417.
5. Stephen Wade. *The Beautiful Music All Around Us. Field Recordings and the American Experience.* (Champaign, Illinois: University of Illinois Press, 2012): 322.
6. Alan Lomax. *The Penguin Book Of American Folk Songs.* (London: Penguin, 1964).
7. Carl Sandberg. *The American Songbag.* New York: Harcourt Brace Jovanovich, 1927: 26 and 217.
8. 'Flying Crow' as sung by Washboard Sam (1910-1966)
9. The Rock Island line closed in 1980.
10. Wade, *Beautiful Music*, 50.
11. Wade, *Beautiful Music*, 48-61.
12. Alan Lomax. *The Penguin Book of American Folk Songs.* (London: Penguin, 1964).
13. *The Washington Post*, September 28, 1903.
14. As cited in Richard Polenberg. *Hear My Sad Story: The True Tales That Inspired "Stagolee," "John Henry," and Other Traditional American Folk Songs.* (New York: Cornell University Press, 2015): 173.
15. Norm Cohen. *Long steel rail: the railroad in American folksong.* (Urbana: University of Illinois Press, 2000): 219.
16. For a full discussion see Cohen, *Long steel rail*, 198 – 218.

Popular music

18 Sounds of the railroad in boogie-woogie, bluegrass, blues and jazz

SEVERAL MUSICOLOGISTS SPECIALIZING in blues and jazz have written about the connection between train sounds and the development of American popular music in the twentieth century.[1] This chapter looks further into these railroad sound connections in the evolution of different popular music styles, boogie-woogie, bluegrass, blues and jazz. The section on jazz focuses on some of the most well-known locomotive-inspired pieces for jazz bands, notably the big band sounds of Duke Ellington and Glenn Miller, moving on to the R&B classic 'Choo choo ch'boogie' by Louis Jordan's band.

The chapter opens with instrumental pieces whose main intention is to replicate the sounds of the train. The favourite instruments for train imitation pieces are the harmonica and the fiddle. The harmonica, or blues harp as it is often known, lends itself to replicating train sounds because of the special effects that can be produced in various ways: cupping the hands over the instrument and opening and closing them to vary the sound; fluttering the fingers; and controlling the flow of the breath. It is also easy to 'bend' notes on the instrument. Note bends, a characteristic of blues and jazz, are where the pitch of a note is changed slightly, usually raised. Each of the following harmonica solos depicting the sounds of a train was recorded in the Southern States of America in the late 1920s.

10 train imitations with harmonica

SONG TITLE	ARTIST
C & O excursion	Frank Hutchinson
C & NW blues	Bert Hunter Bilbro
Chickasaw Special	Noah Lewis
Dixie Flyer blues	De Ford Bailey
Double headed train	Henry Whitter
Lost train blues	Henry Whitter
McAbee's railroad piece	Palmer McAbee
Pan American blues	De Ford Bailey
Railroad blues	Freeman Stowers
Riding the blinds	Eddie Mapp

The two pieces, 'Pan American blues' and 'Dixie Flyer blues', are both performed by the African-American De Ford Bailey who spent many years perfecting the sound of both the engine and the whistle. These impressions are alternated in 'Pan American blues' which imitates the sound of an express freight train. 'Dixie Flyer blues' also alternates the engine and whistle sounds, but it has a more bluesy feel and makes much use of note bends. 'C and NW blues' is performed by the one-time blackface comedian Bert Hunter Bilbro.[2] It opens with an accelerating rhythm imitating the engine sound and then, with great instrumental skill, Bilbro introduces a train whistle sound and actually combines it with the sound of the engine. Palmer McAbee's rendition of 'McAbee's Railroad Piece' is an astonishing virtuoso harmonica solo. McAbee's imaginative interpretation of train sounds opens with the sound of steam escaping as he blows across the top of the instrument. The engine accelerates and when it gets to full speed he uses his throat to produce growling engine sounds.

The fiddle also lends itself to train imitations, this is partly because of double-stopping (where two notes can be played at once) and partly because it is a fretless instrument meaning that it is possible to slide from one note to the next. Most of the pieces listed below are played by string bands which were made up of acoustic stringed instruments - fiddles, five-string banjos, mandolins (occasionally) acoustic guitars and double bass. Apart from the recording of 'Orange Blossom Special', all of the listed pieces were recorded in the 1920s and 1930s when the popularity of the string band was at its height. They are all rooted in bluegrass music, a type of energetic folk music which originated in the American Appalachians and is noted for its rapid tempos and solo improvisations. Country music scholar Charles Wolfe has argued that bluegrass 'resembles the unique sound of the train: the clattering of the drivers echoed by the rolling banjo; the straight, true, hard steel rails resembling the empowering drive of the rhythm guitar; and the wail of the whistle calling up the long, edgy strokes of the low bow fiddle.'[3]

10 train imitations with fiddle

SONG TITLE	ARTIST(S)
C & NW railroad blues	Byron Parker's Mountaineers
Donkey on the railroad track	Al Hopkins And His Hillbillies
Engineer Frank Hawk	Rainey Old Time Band
Lost train blues	Fred Perry (fiddle) Glenn Carver (guitar)
New lost train blues	J E Mainer's Mountaineers
Orange Blossom Special	American Music Shop Band with Mark O'Connor
Peanut Special	Byron Parker's Mountaineers
Southern No 111	Roan Country Ramblers
Train Special	Walter Hurdt and his Singing Cowboys
Western Kentucky Limited	E E Hack String Band

From 1925 to 1953, the *Orange Blossom Special* was a deluxe passenger train on the Seaboard Airline Railroad connecting New York City and Miami. The song 'Orange Blossom Special' was written in 1938 by Ervin T Rouse. It has become best known as a show piece for the fiddle, a vehicle showcasing the player's virtuosity, often performed at breakneck speed. The recording by Mark O'Connor and the American Music Shop Band is particularly impressive, showcasing the pyrotechnic skills of both the fiddler and the harmonica player.

Railroad sounds

The Canadian composer and educator R Murray Schafer, who has written extensively on the soundscape of our environment, writes that in comparison with their modern counterparts, the sounds of steam trains were rich and characteristic.

> Of all the sounds of the Industrial Revolution, those of trains seem across time to have taken on the most attractive sentimental associations…the whistle, the bell, the slow chuffing of the engine at the start, accelerating suddenly as the wheels slipped, then slowing again, the sudden explosion of escaping steam, the squeaking of the wheels, the rattling of the coaches, the clatter of the tracks … these were all memorable noises.[4]

Schafer goes on to write about the sound of train whistles. 'In North America… the whistle is low and powerful, the utterance of a big engine with a heavy load. On the prairies…the periodic whistlings resound like low haunting moans.'[5] Howard Bloomfield, one of Schafer's colleagues on his soundscape projects at Simon Fraser University, believed that the railroads influenced the development of jazz and that 'blue notes can be heard in the wail of the old steam whistles.'[6] By 'blue notes' Schafer was referring to a musical characteristic of blues and jazz where some notes, are flattened by a semitone or 'bent' by a smaller interval. Blue notes are usually found on the third, fifth or seventh degree of the scale. Alan Lomax, arguing that 'the distinctive feeling of American hot music comes from the railroad', also makes reference to the characteristic sound of the whistle. He holds that 'it is the surge and thunder of the steam engine, the ripple of the wheels along the tracks, and the shrill minor-keyed whistles that have colored this new American folk music.'[7]

In his analysis of the ways that the railroads have influenced American popular music, Lomax describes how the influence can be found in the music's texture. The word 'texture' refers to the simultaneously sounding lines in a piece of music, the number of layers, and the way that the rhythms bind together, for example.

> It is in the textures of our popular music, however, that the railroads have left their deepest impression. Listen to the blues, the stomps, the hot music of the last fifty years, since most Americans have come to live within the sound of the railroad. Listen to this music and you'll hear all the smashing, rattling, syncopated rhythms and counter-rhythms of trains of every size and speed. Listen to boogie-woogie with its various kinds of rolling basses… What you hear back of the notes is the drive and thrust and moan of a locomotive.[8]

Boogie-woogie

Lomax makes specific reference to the counter rhythms and the rolling basses of boogie-woogie music. Boogie-woogie is a piano style which was most popular during the 1920s and 1930s but has had a major influence on different rock and blues styles. It is thought to have originated in Texas but became popular in Chicago and Detroit where boogie-woogie pianists provided the music for dancing at 'rent parties' during the years of the Depression. Rent parties were thrown in urban black communities; an entrance fee was charged to pay for food, drink and entertainment and the money was collected to pay the rent. The piano was the only instrument playing so it had to be loud, partly to attract passing custom. The front was taken off and newspaper was put between the strings.[9]

The essence of the boogie-woogie piano style is the contrast between the left and right hands. The left hand plays an ever-recurring driving bass line, sometimes known as a 'walking bass' or a 'rolling bass', keeping the beat and providing the chords. The chords usually followed the twelve-bar blues pattern. At the same time the right hand provides an embellished melody often set up in cross rhythms, what Lomax refers to as counter-rhythms, against the left hand. Here are some examples of the rolling basses of the left hand bass part.

Notice how the note pattern of each bar is repeated, but shifted up or down to the next chord. Boogie-woogie bass lines often use quavers (eighth notes) as in the following example.

Sometimes chords are used providing a useful musical imitation of an engine moving along the tracks.

These are played in a swung rhythm where, in each pair of notes, the first note is performed a little longer than the second and with a little more emphasis.

Honky tonk train blues by Meade Lux Lewis

The boogie-woogie piano piece 'Honky tonk train blues' was first recorded in 1927 and soon became very popular. The chugging engine rhythm can be heard in the right hand. These bars, taken from near the opening, show the polyrhythmic interplay between the left and right hand. The left hand notes come in pairs, whereas those of the right are often in threes – triplets. Boogie-woogie pieces are often technically demanding, mainly because of the rhythmic independence required between the player's hands where twos are set against threes.

In the next section Lewis uses another common device found in the boogie-woogie style where the right hand uses riffs (short repeated figures).

POPULAR MUSIC

No. 29 by Wesley Wallace

This boogie-woogie piece by Wesley Wallace is about the No. 29 train. It is unusual in two ways. In the same way as most popular songs, the very large majority of boogie-woogie pieces have four beats to the bar, whereas 'No. 29' has three beats in a bar. Secondly, Wallace makes a running commentary about his train journey throughout and illustrates this with train sound effects on the piano. The performance is a *tour de force*. The left hand has running quavers throughout depicting the train's motion, whilst the intricate melodic right hand is interspersed by train imitations. Wallace's running commentary is as follows.

This is the train they call 29. Leavin' out of Cairo, comin' to East St. Louis. Soon as she got in Murphysburg, she blowed that whistle. She blowed her whistle this way: (piano imitates whistle with repeated chords in its high register).

I caught that train in Murphysburg. I was intendin' to get off in Sparta, Illinois. I mean, that train was runnin'!

She wasn't doin' nothin' but runnin', hot, a' somethin' like this: (piano imitates train runnin').

I mean that train was runnin'. She wasn't doing nothin' but runnin' hot, something like this

Just before she got to Sparta, she thought she'd blow that whistle again. She blowed that whistle somethin' like this: (piano imitates whistle).

She's lopin' now. I wanted to get off that train, but she's goin' too fast. I hardly ain't touched one foot on the ground, my heel like to knock my brains out. I always step four or five, right tight …and fell off.

This is the noise I made when I hit that ground: (piano imitates impact with clattering chords).

I'm rollin' now. I got up and waved my hand, told 'em, "Good-bye."

This is the way she was cakewalkin' on into East St. Louis: (piano plays in the cakewalk dance style).

Wesley Wallace is not the only musician to set a narrative to music depicting a train. Here are some more examples.

10 Train songs with spoken narratives

Song	Artist
Travelling blues	Blind Willie McTell
The train	Furry Lewis and Will Shade
Talking Casey blues	Mississippi John Hurt
Streamline 'Frisco Limited	Rev. Robert Wilkins
Special streamline	Booker White
Ragtime Texas	Henry Thomas
Panama Limited	Booker White
Going North	Tom Bradford
Frisco leaving Birmingham	George 'Bullet' Williams
Big boy blues	Leon Strickland

Jazz

Duke Ellington and locomotive music

Edward 'Duke' Ellington (1899-1974) was an American pianist, composer and bandleader. He came from a well to do family in Washington DC (his father was a butler at The Whitehouse) and moved to New York in the early 1920s playing at the fashionable Cotton Club in Harlem. This was at the time when big bands were becoming popular and Ellington helped to develop the big band jazz style. His band was made up of three sections – saxophones, brass (trumpets and trombones), and a rhythm section – each playing off each other in a call-and-response style. He wrote hundreds of compositions for them, discovering new tone colours and textures and always writing to the strengths of individual musicians. As well as writing successful popular songs he also composed sophisticated jazz music. Soon his band, with him at the piano, was making tours of the US and then later, worldwide. He continued to lead bands until his death in 1974.

Ellington loved trains and he spent long hours on them travelling on tour. At times he even rented his own train car in order that he and his band members would have a place to eat and sleep when they were touring in segregated towns. The train was his sanctuary. His sister Ruth once told his biographer Derek Jewell that 'You'd see him in a siding somewhere in Texas, the heat at 110, the sweat pouring off him on to a piece of manuscript paper on his knee'. Jewell writes

> He would listen to the chattering of trains at crossings, to the hissing and chuffing as they left stations, and above all the whistles. 'Especially in the South. There the firemen play blues on the engine whistle – big, smeary things like a goddam woman singing in the night'.[10]

He often composed while riding on trains and wrote a whole series of pieces evoking the joys, and often imitating the sounds, of train travel. His locomotive-inspired songs include 'Daybreak Express' 'Happy go lucky local' and 'The old circus train turn-around blues'. His signature tune, 'Take the A train', however was written by Billy Strayhorn. In *Stompin' the Blues*, Albert Murray argues that, although train sounds were integral to such songs, Ellington's compositional approach was in terms of form and craft rather than 'railroad mimicry'. Murray writes 'although the railroad sounds in such Ellington compositions...remain unmistakable...even the most literal imitation of the sound of the most familiar everyday phenomenon becomes an element of musical stylization and convention... what all the whistles, steam-driven pistons, bells, and echoes add up to is the long-since-traditional sound of blues-idiom dance-hall music.'[11] This point is reinforced by Jewell when he writes that the 'feeling for trains is obvious; but what Duke added to the mixture of train-like sounds and rhythms provided the magic'.[12]

10 train pieces by Duke Ellington

Take the A Train
Daybreak Express
Across the track blues
Lightnin'
The old circus train turn-around blues
Choo choo (gotta hurry home)
Build that railroad (sing that song)
Loco Madi
Happy go lucky local
Track 360 aka Trains that pass in the night

'Daybreak Express'

Barney Bigard, a clarinettist in Ellington's band once said in interview

> You know the record, 'Daybreak Express'? Well, when we were in the South, we'd travel by train in two Pullmans and a baggage car. Duke would lie there resting, and listening to the trains. Those southern engineers could pull a whistle like nobody's business. He would hear how the train clatter over the crossing, and he'd get up and listen to the engine. He'd listen as it pulled out of a station, huffing and puffing, and he'd start building from there... He had the whistles down perfectly, too.[13]

'Daybreak Express' (1933) is built on the chords of 'Tiger Rag', an earlier Ellington song. It portrays a steam engine leaving the station, accelerating to top speed, blowing its whistle and then coming to a halt. It provides one of the most vivid pictures of a speeding train in music. With its noisy, powerful, chords, rolling ostinati, complete with wailing whistle and clanging bell, it has a sense of the immediate urgency of a pounding express. The piece abounds with bends and slides on the saxophones and wah-wah trumpets. In 2013 the *Los Angeles Times* maintained that 'Daybreak Express' was rarely performed 'because no one has figured out precisely how he got his reeds to replicate the sounds of a steam whistle.'[14]

'Happy go lucky local'

In his autobiography, Ellington wrote that 'Happy go lucky local'

> told the story of a train in the South, not one of those luxurious, streamlined trains that take tourists to Miami, but a little train with an upright engine that was never fast, never on schedule, and never made stops at any place you ever heard about. After grunting, groaning, and jerking, it finally settled down to a steady medium tempo.[15]

The big band composition opens with rumbles, groans, shrieks, squeaks and dissonance, one repeated pattern (ostinato) after another, all set over a steady beat. There is no clearly identifiable melody until well into the piece. In the early bars we hear piano and bass as the train starts to move. Reeds and brass enter with saxophone solos and long, screaming notes on plunger muted trumpet, reminiscent of the whistle of a passing train, all accompanied by the trombones, sounding like the chugging of the train's engine.

'Take the A train'

'Take the A Train' is a jazz standard, composed in 1939 by Billy Strayhorn. Ellington was in the middle of a negotiating stalemate with the licensing agency American Society of Composers, Authors and Publishers (ASCAP); all music by ASCAP members—including Ellington himself—was barred from the airwaves. Ellington needed a new library of songs by a non-agency member, and turned to his young friend – the arranger and collaborator, Billy Strayhorn. Its opening theme is unusual in its use of wide leaps, but it is also very hummable. Ellington soon adopted it as his signature tune. The lyrics are about choosing the then-new A train to get to Harlem in preference to the D train. Apparently a common mistake was to take the 'D' train and end up in the Bronx. The story goes that the title evolved from the directions that Ellington gave Strayhorn on how to get to his Harlem apartment by subway. Ellington lived in the elegant African-American of Sugar Hill located between 144th and 155th Street. As Strayhorn once said, he was writing subway directions.[16]

You must take the A train
To go to Sugar Hill way up in Harlem
If you miss the A train
You'll find you've missed the quickest way to Harlem
Hurry, get on, now it's comin'
Listen to those rails a-hummin'
All aboard get on the A train
To go to Sugar Hill way up in Harlem.

The song is in the 32-bar AABA form which was popular with swing and jazz bands of the time. It opens with solo piano playing the same bar of downward flourishes four times. The locomotive is suggested in the constant movement and rolling ostinato of the rhythmic brass and reed accompaniment. The joyful main theme is first played by saxes in unison, punctuated by the other horns. A muted-trumpet solo backed by the reeds leads to another stirring riff by the full band. A long trumpet solo follows, first muted and then open. The song changes key by means of blasts from the full band in the passage between the muted and open trumpet. Loud dissonant chords from the band towards the end of the solo are suggestive of a train's warning signal.

Ellington and the band performed the song in the film *Reveille with Beverly* (1943) with the singer Betty Roche. The band is shown performing in a railroad passenger car rather than a subway car. The song was an instant hit and since Ellington's recording there have been many others including those by Ella Fitzgerald, Joe Henderson, Charlie Mingus, Sun Ra, and the rock band Chicago.

The Glenn Miller Orchestra

Glenn Miller (1904-1944) was a bandleader, trombonist, composer and arranger. His band had a more commercial sound than Duke Ellington's. In his search for a unique sound he doubled he clarinet and tenor saxophone, whilst three other saxophones played in close harmony. He was the best-selling recording artist from 1939 to 1942 with 16 number-one records and 69 Top Ten hits. In 1942, he volunteered to join the US military to entertain World War II troops. He joined the US Army Air Forces, directing the American Air Force band and contributing a good deal to morale during the war. On December 15, 1944, when France had been newly liberated, Miller's Paris-bound plane disappeared in bad weather over the English Channel. Neither the plane nor its occupants were ever found.

'Chattanooga Choo Choo' was written by Harry Warren and Mack Gordon for the band to perform in the film *Sun Valley Serenade* (1941). It was arranged for the Glenn Miller Orchestra by Jerry Gray. The band were incorporated into the story line, the lyrics describe the train's route, starting from Pennsylvania Station in New York and running through Baltimore to Carolina before reaching Chattanooga. A past love will be waiting at the station and he plans to settle down for good with her. At the time soldiers abroad were facing death, the song reminded them of the excitement of coming home. The main song includes a passage of dialogue.

"Pardon me, boy, is that the Chattanooga Choo Choo?"
"Yes, yes, Track 29!"
"Boy, you can give me a shine."
"Can you afford to board the Chattanooga Choo Choo?"
"I've got my fare, and just a trifle to spare."

The train station in Chattanooga, Terminal Station opened in 1908 and was a major transportation hub in the 1930s and 1940s. Gordon and Warren were reputed to have written the song while traveling on the Southern Railway's *Birmingham Special* train - one of three trains operating from New York City via Chattanooga. However some of the details in the song don't apply to the *Birmingham Special*, suggesting a certain amount of artistic licence on the part of the team. The song mentions 'Track 29' but New York's Pennsylvania Station only had 21 tracks. The *Birmingham Special* left at 12.

30 rather than 'bout a quarter to four'. Although 'nothing could be finer' than to eat 'dinner in the diner' it would be difficult to do this since none of the trains between New York City via Chattanooga passed through Carolina. Chattanooga's former Terminal Station was saved from demolition after the withdrawal of the passenger rail service in the early 1970s. It is now part of a large resort complex and Chattanooga is home to the National Model Railroad Association.

That year it won an Academy Award for Best Song from a movie, even though it could not be heard on network radio due to the ASCAP boycott, the same boycott which had brought about the composition of 'Take the A Train'. Once the ASCAP strike was over it was featured heavily on radios across the country and in 1942 it went on to receive the first Gold Record for selling over a million copies. Other railroad pieces performed by the Glenn Miller Orchestra were 'Tuxedo Junction', Sleepy Town train', 'Slow freight' and the Count Basie song '9.20 Special'.

10 choo choo songs

SONG TITLE	ARTISTS
Boogie-woogie choo choo train	Mabel Scott 1948 R&B
Bye bye black smoke choo choo	Recordings by bluegrass artists Joe Glazer, The New Lost City Ramblers, and Arthur Smith and his Crackerjacks (1955)
Chattanooga choo choo	Glenn Miller and his Orchestra 1941 Big band
Choo choo	Frankie Trumbauer 1930 Big band
Choo choo	Jack Hylton 1931 Big band
Choo choo	Fats Waller 1939 Piano
Choo choo (gotta hurry home)	Duke Ellington 1924 Jazz band
Choo choo blues	Virginians 1922 Blues band
Choo choo ch'boogie	Louis Jordan 1946 R&B
When the midnight choo-choo leaves for Alabam'	Written by Irving Berlin in 1912. Recordings include Judy Garland, Ethel Merman and Tommy Dorsey

'Choo choo ch'boogie' by Louis Jordan & His Tympany Five

Louis Jordan (1908 –1975) was an American saxophonist, bandleader and songwriter whose music was so popular that he was sometimes referred to as the King of the Jukebox. Born in Arkansas, Jordan left home in his teens and eventually moved to New York and formed the Tympany Five, an R&B dance band made up of a horn section along with drums, double bass, guitar and piano – a sort of scaled down big band. R&B (or rhythm and blues) was a blues style made for entertainment and dancing. Jordan teamed up with Milt Gabler, Decca's producer of 'race' records to produce a series of hits and he became the first black artist whose records crossed over to the pop charts.

The song was co-written by Milt Gabler and two country and western musicians who worked at a radio station in New York City. Like many of Jordan's songs 'Choo choo ch'boogie' is in the jump blues style, a sort of swing-oriented happy version of the blues with a boogie-woogie beat. The verses follow the 12-bar blues structure and the choruses have eight bars. It has rapid-fire witty lyrics and a hard-driving energetic sound. The record was released in 1946 and was soon a massive hit selling over two million copies. Part of its appeal is the way that the lyrics encapsulated the feelings of excitement, coupled with disillusionment, which many must have felt on returning from service in World War II; 'The only job that's open needs a man with a knack'.

The song opens with a 12-bar instrumental introduction in which the horns (two saxophones and a trumpet) imitate the sound of a train whistle. The rhythm section sets up a boogie-woogie shuffle and the vocals are joined by riffing horns, a boogie-woogie piano solo and a 20-bar saxophone solo. The rapidly fired vocals have their emphasis on the rhythm and some, including those of the catchy chorus, are onomatopoeic – 'whoo, whoo', 'choo, choo' - suggesting the sounds of a train.

The success of the song inspired a number of the first rock 'n' roll artists including Bill Haley and his Comets.

Endnotes

1. These include the blues expert, Paul Oliver, American folklorist, Alan Lomax, country music expert Charles Wolfe, and the American jazz and blues critic, Albert Murray.
2. White performers blackened their face with cork for the performance of black caricatures in minstrel shows.
3. Taken from the sleeve notes to the album *Bluegrass Express*, a compilation of train themed bluegrass songs.
4. R Murray Schafer. *The soundscape. Our sonic environment and the tuning of the world.* (Rochester, Vermont: Destiny Books, 1977): 81.
5. Schafer, *The soundscape*, 82.
6. Schafer, *The soundscape*, 113.
7. Alan Lomax, ed. *Folk Song*: USA, 1947.
8. Ibid.
9. Eileen Southern. *The Music of Black Americans. A History.* (New York: Norton, 1971): 372.
10. Derek Jewell. *A Portrait of Duke Ellington.* (London: Sphere Books, 1977): 79.
11. Albert Murray. *Stomping the Blues.* (Minneapolis: University of Minnesota Press, 1976): 125.
12. Jewell, *A Portrait of Duke Ellington*, 79.
13. James Lincoln Collier. *Duke Ellington.* (Oxford: Oxford University Press, 1987): 162.
14. *Los Angeles Times*, May 10, 2013.
15. Edward Kennedy Ellington. *Music is my mistress.* (New York: Doubleday. 1974): 162.
16. Jewell, *A Portrait of Duke Ellington*, 89.

19 A medley of popular songs

RAILWAYS AND TRAINS are so deeply ingrained in the popular imagination that they feature in hundreds, possibly thousands of popular songs. Consequently, it is difficult to make a selection. The featured songs are not intended to represent the best songs, but most have interesting stories behind them; some are included because they are so well-known, such as the opening song 'The Loco-Motion', others because they are personal favourites, 'Click clack' by Captain Beefheart and Hugh Masekela's 'Coal train', for example. Individual songs are grouped under common themes, and partly to acknowledge the arbitrariness of the overall selection, some of these appear in lists - last trains, fast trains, slow trains, trains leaving and trains heading home. This section covers the two pop song classics 'Midnight train to Georgia' and 'Homeward bound' with an attempt to establish whether there is any truth in the much-repeated idea that Paul Simon wrote the latter on the platform of Widnes railway station.

In some songs, train references are confined to the lyrics and no attempt is made to convey the sounds of the railway. In other songs, train imitations are at the forefront, and a group of these come under the heading 'The sounds of the train in late 1960s and 1970s albums'. Featured artists include David Bowie, Tangerine Dream frontman Edgar Froese, and Kevin Ayers, all taking advantage of what were then recent advances in music technology.

There are three American artists in particular who have made a surprisingly large number of references to railroads in their songs. These include Tom Waits who is, in his own words, obsessed with train sounds, and Grateful Dead, whose repertoire includes at least 17 railroad songs, most famously their own version of 'Casey Jones'. This song is explored in more detail, in particular the notion that Casey Jones was high on cocaine when his train crashed. Bob

Dylan's first recording of a train song dates from 1962 when he included 'Freight train blues' on his debut album and his most recent is 'Duquesne whistle' from his 2012 album *Tempest*.

Three very well-known songs for children form another group: 'I'm a train', 'The runaway train' and 'Morningtown ride'. The chapter then moves on to train journeys from around the world, a disparate mix of places and styles ranging from Jamaica to South Africa, and the Cuban beat boxing of Vocal Sampling to the German electronica of Kraftwerk. The final chapter closes by exploring music from the twenty-first century, questioning whether anyone still writes popular songs about trains and, if they do, are any of them in one of the styles which currently predominate the charts.

This medley of popular of songs opens with two of the most famous popular songs about trains, one fast and one slow.

The Loco-Motion by Little Eva

'The Loco-Motion', composed by Gerry Goffin and Carole King, has been a hit several times. On its first release, sung by Little Eva, it became an international hit in 1962. It was next in the US charts in 1974 performed by the rock band Grand Funk Railroad, and finally it was a No. 1 hit in Australia for Kylie Minogue in 1988. Goffin and King were a husband and wife songwriting partnership who worked in the famous Brill Building – a New York 'hit factory' of the time. The pair had many well-crafted hits together and wished to capitalise on the dance craze of the early 1960s in which dances were pre-packaged with their own theme song, perhaps most famously in 'The Twist' by Chubby Checker. In the search for a singer they looked no further than their teenage domestic help, Eva Narcissus Boyd (1943-2003) aka Little Eva. Eva Boyd was not the inexperienced singer she is sometimes presented as; she had performed in the family gospel group and whilst studying in New York had worked as a session singer with the girl-group Cookies. Carole King was impressed by her strong, bright voice and used her in the recording with backing vocals from the Cookies and King herself on keyboards. Sustained drone notes on the saxophone in the song's opening help to evoke the sound of a steam whistle. The lyrics exhort the listener to take part in the latest dance craze, to do the locomotion, chugging like a railroad train. Given that up to this point the locomotion dance did not exist, one had to be invented and Eva Boyd was given the task of creating a dance to fit the song with train-like movements. With its dense texture combining the double-tracked vocal of Little Eva, rhythmic layers of drums, hand claps, saxophone, vocals and female backing singers, the overall 'big' sound is not unlike Phil Spector's Wall of Sound.

Slow train by Flanders and Swann

Swathes of the rail network were closed as a result of the 1963 government report *The Reshaping of British Railways* under the chairmanship of Richard Beeching. The report was targeted at the removal of underused and unprofitable lines. Between 1965 and 1968, 2695 stations had closed and over 2000 miles of line were no longer used.[1] Small branch lines and rural stations were particularly badly hurt. The Beeching Cuts, as they are often referred to, are frequently criticized in terms of the consequent reduction of the number of rural train services and of the loss of the hundreds of small stations that had played an important role in community life. During the nineteenth century, steam trains had been perceived as the epitome of modernity, but this perception eventually faded and instead steam trains became a symbol of nostalgia 'coloured by a sense that something more fundamental than a transport service was lost when railways closed'.[2]

10 songs about fast and slow trains

SONG	ARTIST(S)	SONGWRITER
Fast freight	The Kingston Trio, Serendipity Singers, Ritchie Valens	Terry Gilkyson
Fast freight	Ritchie Valens	Ritchie Valens
Fast movin' train	Restless Heart	Dave Loggins
Fast train	Van Morrison, Solomon Burke	Van Morrison
Fast train through Arkansas	The Delmore Brothers, Wayne Raney	Alton and Rabon Delmore
I been to Georgia on a fast train	Billy Joe Shaver, Johnny Cash, Willie Nelson	Billy Joe Shaver

Slow train	Flanders and Swann	Flanders and Swann
Slow train	Bob Dylan	Bob Dylan
Slow train to nowhere	John Mayall	John Mayall
Woody and Dutch on the slow train to Peking	Rickie Lee Jones	Rickie Lee Jones

Flanders and Swann's melancholic song 'Slow train', published in 1963, is a paean to the imminent closure of many small stations. An elegiac list of bucolic place names is intoned over a simple lilting piano accompaniment - Blandford Forum, Buttermere, Midsomer Norton and Tumby Woodside. A gentle and atmospheric song, it ponders the unhurried delights of a slow train, a place to meet where the stations have 'whitewashed pebbles', the 'grass grows high' and the 'sleepers sleep', a place where you might find a 'cat on a seat'. At the same time the song features some of the elements, such as over-staffing with porters, that Beeching was keen to eliminate. It is a poignant, evocative song, a lament for the passing of a way of life.

Trains leaving and trains coming home

Train journeys are often linked with notions of connection and isolation, proximity and distance, presence and absence, leaving and coming home. They can separate families and lovers and they can bring them back together again. Two hit songs from the 1960s put these feelings of loss on separation unequivocally. In 'Homeward bound' the absent lover Paul Simon puts it succinctly in that familiar line, 'I wish I was homeward bound', and in the lyrics of the Burt Bacharach song 'Trains and boats and planes', we are told that trains can take you away and bring you back home, giving the means of transportation the responsibility for the coming and going rather than the passengers who can only pray for the return of their loved ones.

10 songs about trains heading home

SONG	ARTIST(S)	SONGWRITER
First train home	Fleetwood Mac	Peter Green
Going home train	Lawrence Winters	Harold Rome
Home in a boxcar	Hoots & Hellmouth	Sean Hoots
Homebound train	Bon Jovi	Jon Bon Jovi, Richie Sambora
Homeward bound	Paul Simon	Simon and Garfunkel
I'll be home on Christmas Day	Elvis Presley	Michael Jarrett
Last train home	Pat Metheny	Pat Metheny
Train fare home	Muddy Waters	Muddy Waters
Train home	Rich Moore & Mollie O'Brien, Chris Smither	Chris Smither, Patty Larkin
Train ride home	lofi.samurai	lofi.samurai

Homeward bound by Paul Simon

In an attempt to launch a solo career in England, the American singer-songwriter Paul Simon embarked on a short tour of English folk clubs in 1965. In this autobiographical lament he sings of his displaced lifestyle, living out of suitcases and getting on trains each day to travel to the next venue. Whilst in England he fell in love with a young woman, Kathleen (Kathy) Chitty, who took the ticket money on the door of the Hermit Folk Club in Brentwood, and this song is about her. 'Homeward bound' was released on the Simon and Garfunkel album *Parsley, Sage, Rosemary and Thyme* in 1966. The style could be described as soft rock, somewhere between folk music and rock, with listenable melodies and vocal harmonies. The song opens with a short acoustic guitar solo moving into the first verse. It has a straightforward verse and chorus structure in which the verses tell of the singer's disillusionment with his lonely touring lifestyle and the lyrics of the chorus express his longing to be heading home to his love. Simon uses the lower register of his voice as he details his day-to-day life as a musician on the road and then in the chorus moves into a higher register when describing his yearning to be home. At the end of the final chorus, the words 'silently for me' are repeated, the instruments fall away and the song concludes with the acoustic guitar figure of the opening.

It is often reported that Paul Simon wrote 'Homeward bound' whilst waiting for a train on Widnes railway station platform.[3 4] Although a plaque was erected there in 1990 to mark the site, as far as can be established, this story is almost certainly untrue. What is fairly certain is that Simon was undertaking a tour of Northern folk clubs when he wrote the song and that he performed in Widnes on September 13, 1965.[5] It is unclear what his destination was. Some reports say Manchester, others Liverpool and yet others London, so no clues can be found from railway routes and timetables.

What does Paul Simon have to say about it himself? In 2000 Lyn Goldsmith interviewed him for *The Times* and asked him whether he had written the song on Widnes station. His reply was "Well, no, not actually in the station, but around that time. While I was up that way".[6] In his biography of Simon, Robert Hilburn writes that the singer was clear that the song 'grew out of the Northern England tour, even specifying on occasion the time he sat in a railway station in Widnes'. Hilburn, however, gives the final word to Geoff Speed who 'ran the Windsor folk club in Widnes'[7] and 'drove Simon to the Widnes train station the day he was supposed to have written the song'.

Widnes Station (formerly Widnes North) - station building on south platform - May 1992
© Eddie Hewison.

> "It has always been a sweet story, but there's no way he could have written the song at the station," Speed said. "The thing I remember most about the morning was that we got to the station just as the train pulled in, and Paul had to run to make it. He didn't have time to sit down, much less write a song."[8]

On the other hand, Speed is quoted as saying elsewhere

> It is probable he wrote one verse in Liverpool and the chorus in Wigan, with the song being finished in Widnes. We heard him writing the tune when he was staying at our house and then we dropped him at the station. He probably finished the song on the platform.[9]

Given the lack of hard evidence to support the story, why does it still persist? Hilburn writes that it was encouraged by the local population whose town was more used to negative portrayals.

> In Widnes the story was hard to kill… Welcoming any positive attention – the town was long the butt of jokes because of an awful smell caused by fumes from local chemical plants – townsfolk took pride in and placed a plaque outside the train station.[10]

So bad was the smell in Widnes that the railway historian Simon Bradley writes that 'Rails wore out faster…in places such as Widnes, where heavy industry made the very air chemically corrosive'.[11]

The music critic Richard Morrison writing in *The Times* observes that it is 'strange how often the most gripping literature, music and painting is created in grim circumstances' and goes on to cite the Widnes station story.

> Paul Simon allegedly composed one of his most celebrated songs 'Homeward Bound' while shivering on the platform of Widnes railway station – an unlikely conjunction of inspiration and circumstance now noted for posterity on a wall plaque. "If you've ever been to Widnes," Simon said, "you'll know why I was so desperate to be homeward bound".[12]

And, I would argue, that it is this incongruous juxtaposition of grim Northern town and American pop classic which perpetuates what could well be an urban myth.

10 songs about trains leaving

SONG	ARTIST(S)	SONGWRITER
Desperados waiting for a train	Nanci Griffith, Tom Rush, Martin Simpson, Willie Nelson, Johnny Cash	Guy Clark
Don't miss that train	Sister Wynona Carr	Sister Wynona Carr
Just missed the train	Kelly Clarkson, Carly Hennessy, Trine Rein	Danielle Brisebois, Scott Cutler
Midnight Train to Georgia	Gladys Knight and the Pips	Jim Weatherley
Stop that train	Bob Marley, Peter Tosh, Jerry Garcia Band, Party Animals	Peter Tosh
Stop that train	Spanish Town Skabeats, The Workingmen featuring Sly & Robbie	Prince Buster
Waitin' for the train to come in	Maria Muldaur, Peggy Lee, Louis Prima	Sunny Skylar, Martin Block
Waiting for a train	Jimmie Rodgers, Johnny Cash, Jerry Lee Lewis, Billy Bragg	Jimmie Rodgers
Waiting for the '103'	Dan Hicks and His Hot Licks	Dan Hicks
Wave the flag and stop the train	The Move	Roy Wood

Midnight Train to Georgia by Gladys Knight and the Pips

In 1973, Gladys Knight and the Pips topped the US charts with a song with a recurring line about a train leaving, bound for Georgia. Much has been written in previous chapters about the mass migration from the land in the Southern States to cities further north in search of a better standard of living. This train, however, is travelling in the opposite direction, reflecting the lives of many migrants who found that displacement to cities such as Detroit, Chicago and Los Angeles had left them with a nostalgic desire to return to a life down home. Composed by Jim Weatherley, the lyrics are written from the perspective of a woman whose lover had come to Los Angeles dreaming of stardom, but it had proved too much: he became disillusioned and was leaving on the midnight train to Georgia.[13] It is a heartfelt soul song with the distinctive sound of Tamla Motown - strings, horn section and drums - although Gladys Knight no longer recorded with Tamla at this point.

Stop that train by Bob Marley / Peter Tosh[14]

As we have seen in previous chapters, some trains held the promise of escape from hardship. In his lyrics to the Bob Marley song 'Stop that train', Peter Tosh draws heavily on this theme of escape so often found in blues and gospel songs. It is a song of hopelessness and loneliness, 'People are scattered, misdirected'; some are 'living successfully, but most...are struggling and starving'. The song has all the musical hallmarks of Jamaican reggae. As Kwame Dawes puts it in his analysis of lyrics in Bob Marley's songs, this 'lamentation' is 'contradicted by the danceability of the song'.[15]

The earlier ska song 'Stop that train', recorded by the Spanish Town Skabeats in 1965, has a simpler message; a woman whose boyfriend is leaving cries "stop that train" because she wants to join him on board.

Another common theme is that of the last train. Sometimes the train has a specific destination, such as London or Clarksville or, as in Arlo Guthrie's song 'Last train', it is headed for glory. At other times the last train is a specific type, often a steam train, and there are also several examples where the 'last train' is unspecified; rather, it acts as a metaphor.

20 songs about last trains

SONG	ARTIST(S)
I took the last train	David Gates
Last chance train	Bon Jovi
Last of the steam-powered trains	The Kinks
Last steam engine train	Leo Kottke
Last train	Yes
Last train	Primal Scream
Last train	Arlo Guthrie
Last train	Allen Toussaint
Last train	The Backsliders
Last train	Dead Moon
Last train	Graham Central Station
Last train	Jerry Reed
Last train	Captain Sensible
Last train	Travis
Last train home	John Mayer
Last train home	Blink-182
Last train home	Lostprophets
Last train to Awesometown	Parry Gripp
Last train to Clarksville	The Monkees
Last train to London	Electric Light Orchestra

Last train to Clarksville by The Monkees

The Monkees were an American band that was put together in 1966 for a television series of the same name. 'Last train to Clarksville' was the band's debut single which, to everyone's surprise, topped the American charts. A catchy Beatles-inspired pop song, it was written by Tommy Boyce and Bob Hart and recorded by session musicians. Hart wished to capture some of the Beatles' feel and sound, so, for example, the 'Oh, no, no, no' line was a response to the Beatles' 'yeah, yeah, yeah' lyrics. The lyrics tell of a man calling his girlfriend, urging her to meet him at the Clarksville railway station before he leaves. 'Last train to Clarksville' was written at the height of the Vietnam war; hence some have speculated that the it may have been an anti-war song, particularly because of the words of the final verse where the singer wonders if he will ever be coming home, perhaps a soldier leaving for war. There has been some debate as to which Clarksville the song refers to, given that there are several places with that name in the US. The theory has been put forward that it is Clarksville, Tennessee, where there was an air base, the home of the 101st Airborne Division which was then serving in Vietnam. This idea would tie in with the potential anti-war theme, but it has been denied by the lyricist Bob Hart, who said the place name came from tweaking Clarkdale, Arizona, which he passed through often on his summer holidays.

Last train by Captain Sensible

The opening lyrics of Captain Sensible's 1995 song 'Last train' refer to Dr Beeching chopping up the branches. The Damned's co-founder ex-punk Captain Sensible is a keen train enthusiast and has had a Class 47 diesel locomotive, 47810, named after him by Cotswold Rail. In a 2009 interview he revealed that 'The Damned use touring to pursue our train obsessions' and visit 'as many steam preserved lines as possible' when out on the road.[16]

Last of the steam-powered trains by the Kinks

In 1968 the last steam-driven passenger train was withdrawn from the British railway. Later that year the Kinks released their song 'Last of the steam-powered trains' on their album *Village Green Preservation Society*, a collection of vignettes of pastoral English life. The song is a lament for the passing of steam, at the same time acting as a metaphor for the past and how to deal with it. The singer compares himself with the 'good old fashioned steam-powered trains', the 'last of the blood and sweat brigade' and the 'good old renegades'. The Kinks song, written by their singer Ray Davies, was inspired by Howlin' Wolf's Chicago blues classic 'Smokestack lightnin', which is in itself a train song, 'smokestack lightning' referring to the sparks that fly out of a locomotive's smokestack. The two songs are indeed very similar, both making much use of the Howling Wolf riff heard throughout 'Smokestack lightnin' and both with very simple harmony and intermittent blasts of harmonica.

The sounds of the train in late 1960s and 1970s albums

For some artists, the late 1960s and the 1970s were a period of experimentation in pop music. Technology had opened up more advanced recording techniques: stereo sounds were being exploited more widely and electronic instruments, samplers and synthesisers, had been introduced. It was also the heyday of psychedelia when the 'mind-expanding' hallucinogenic properties of drugs were expressed in music.

Stop this train (again doing it) by Kevin Ayers

'Stop this train' (1969) is a track on *Joy of a Toy*, the debut solo album of Kevin Ayers (1944-2013) a founding member of the underground band Soft Machine. The album features two of Ayers' erstwhile Soft Machine colleagues, Robert Wyatt (drums) and Mike Ratledge (organ), along with the composer David Bedford on piano and Mellotron. Bedford was also responsible for the song's instrumentation and arrangement. The haunting lyrics of 'Stop this train' tell of a nightmare journey by a 'train to anywhere', where it 'don't stop for anyone' and it is impossible to get off. Piano and guitar open the song emulating the sound of a train gaining momentum. As the tempo increases, so does the speed of the recording, taking us into the realms of psychedelia. Bedford was a musical polymath and a leader in the field of music education, specialising in work which did not use musical notation. During this period *musique concrete* techniques, distorting taped sounds by editing techniques such as reversal and speed changing, were used by Bedford and others in schools. The speeded-up tape effect of 'Stop this train' then blends into a more normalised strumming guitar and drums pattern. Once the train is on its way, the drum rhythm stays the same for the full six minutes of the track, adding to the relentless nature of the journey. The nine verses of the song are interrupted twice by instrumental interludes overlaid with train whistle sounds from Mellotron and guitar. In the second half of the song there is passage of swirling duetting between Ratledge and Bedford, intensified by stereo effects, gradually getting wilder until the music sounds almost out of control and fades out.

Station to Station by David Bowie

David Bowie's 1976 album *Station to Station* opens with over a minute of the sound of a train approaching, rebounding from speaker to speaker. The train sound effects on 'Station to Station' were produced by the guitarist Earl Slick using flangers, feedback and delay effects.[17] The sound is panned across the stereo speakers before it fades out, almost as though it is disappearing into a tunnel.[18] The somewhat obscure lyrics of 'Station to Station' introduce the next of Bowie's ever-changing images, the sinister persona, the Thin White Duke. No further reference is made to trains or railway stations in the lyrics of the song. According to Nicholas Pegg, a leading authority on the life and work of Bowie, the title instead refers to the Stations of the Cross.[19] Pegg also holds that Bowie was influenced by German electronica bands. The first track of the album *Epsilon in Malaysian Pale* (1975) by Edgar Froese (frontman of Tangerine Dream) opens with the sounds of a travelling train overlaid by jungle noises. The train sounds were produced using an analogue synthesiser, Mellotron and flute. *Epsilon in Malaysian Pale* was a likely influence on Bowie as was another German electronic band, Kraftwerk. Kraftwerk's pioneering album *Autobahn* (1974) begins with the sound of a car panning across the stereo speakers.[20] In 1977 Kraftwerk returned the compliment on their album *Trans-Europe Express* where they make reference to Bowie and *Station to Station* (see page 273-4).[21]

Click clack by Captain Beefheart

The music of Captain Beefheart (1941-2010) and his Magic Band combines elements of the blues with experimental rock and free jazz, and the lyrics are opaque with a Dadaist weirdness, both poetic and nonsensical. Beefheart's music is incomprehensible to some and a work of genius to others; many musicians, including Tom Waits, Nick Cave, Oasis, and the Red Hot Chilli Peppers, have cited him as an influence. Don Van Vliet was born in Glendale, California, where he discovered the blues with his childhood friend Frank Zappa. Together they came up with the name Captain Beefheart, and Van Vliet formed a blues band, the Magic Band, in the mid-1960s. Over the years the Magic Band has had shifting personnel with some highly accomplished musicians, notably the drummer Drumbo (John French) and guitarist Zoot Horn Rollo (Bill Harkleroad). Many fans regard the band's masterpiece to be the 1969 album *Trout Mask Replica*, which was produced by Frank Zappa. Beefheart retired from the music business in 1982 and became reclusive, returning to his home near the Mojave Desert in southern California, where he reverted to his real name and pursued a successful career as an abstract artist.

In the onomatopoeically titled song 'Click clack' (1972), Captain Beefheart sings of his girl who is going to New Orleans to get herself 'lost and found'. There are two trains and two railroad tracks, one going and one coming back. 'Click clack' makes much use of repeated train-derived rhythms piled up on top of each other in layers. This is a rhythmic device used in many other train songs, but Beefheart's approach is different. At first hearing the interlocking rhythms may appear to be random but they are not improvised and have a logical foundation. The song opens with a three-note figure on the bass and piano, but when the drums motor in with the first locomotive pattern it is evident that the downbeat is not where it first appears to be. Rhythmic patterns appear for a few bars and then change, sometimes the time signature changes too, starting with three in a bar and then four, and shifting the metre throughout. These train rhythms are overlaid with steam whistle sounds, slide patterns on the guitar and harmonica runs weaving in and out. 'Click clack' moves along at a ferocious pace until it reaches the final words where the train leaves and the girl makes her farewell as she waves her handkerchief.

Bellerin' Plain by Captain Beefheart

'Bellerin' Plain' (1970) uses a similarly complex rhythmic technique with layers of interlocking repeated patterns. The song opens with the words 'Parapliers[22] the willow dipped', a phrase which was later used as the title for an exhibition of Van Vliet's abstract art. The lyrics are suffused with railroad imagery (tracks and smokestacks, the 'fireman 'n the brakeman', the cowcatcher and the engineer), and with phrases such as roots 'gnarled like rakers' and 'steel flash scream', the words are reminiscent of the surreal poetry of Lewis Carroll. In 'Bellerin' Plain', much use is made of the marimba, an instrument similar to the xylophone and rarely found in pop music. Halfway through the song there are passages where the marimba duets with the guitar, locked in fugal counterpoint. The song ends with a wild soprano saxophone solo played by the Captain himself.

Some artists stand out as being particularly fond of featuring trains in their songs. These include Bob Dylan, Tom Waits and the Grateful Dead.

Train references in the songs of Tom Waits, the Grateful Dead and Bob Dylan

Trains are found everywhere in the music of Tom Waits, not just in the lyrics, where they often linked to escape, homecoming or nostalgia, but also in their locomotive sounds in songs such as 'Clang boom steam'. In their Tom Waits blog, Sam Whiles and Tom Kweil list 28 songs where trains feature. These include four songs which have the word 'train' in the title – 'Gospel Train', 'Downtown train' (covered by Rod Stewart and others), 'Train song' and 'Down there by the train'.[23] In a 1992 interview, Tom Waits said, 'I've got a lot of trains on tape. Real chugs that are like a rhythmic chug... And the ting ting ting as the bell's coming up.'[24] He talks about his Chamberlain 2000 synthesiser 'It's got a variety of trains; it's a sound that I've become obsessed with, getting an orchestra to sound like a train... I have a guy in Los Angeles who collected not only the sound of the Stinson band organ… but he also has pitched four octaves of train whistles so that I can play the train whistle organ, which sounds like a calliope'.[25]

Ken Rattenne in his essay 'The railroad as metaphor' lists 17 Grateful Dead songs containing railroad references, pointing out that the list is not conclusive and does not include all their cover songs. Amongst the songs listed are 'Caution (do not stop on the tracks)', 'Terrapin station', 'Tons of steel', 'New potato caboose', 'Jack Straw', and their most well-known railway song, 'Casey Jones', written by band member Jerry Garcia in 1969 (for more information about the legendary hero Casey Jones and the original song see pages 210-213). The opening words of the Grateful dead version of the legend claim that Jones was speeding along, driving a train when 'high on cocaine'.[26] Jones' use of cocaine is more likely than it at first might appear. Research has shown that the first epidemic of cocaine use in America occurred during the late 19th century when there were no laws restricting its sale or consumption and it was freely available in drug stores, saloons, and even grocery stores.[27] Until 1900 Coca-Cola contained small quantities of cocaine and in 1993 *The Journal of Clinical Pharmacology* reported that 'Bartenders often added punch to their drinks by adding a pinch of cocaine to whiskey' and that 'Some employers in the construction and mining industries even distributed cocaine to their workers to keep them going at a "high pitch".[28] Other references to cocaine are made in the chorus of 'Casey Jones', which includes the punning line 'Watch your speed'; however, at its moderate pace the song is rather a stately train journey for a hard rock band.

The lyrics to 'Casey Jones' were written by Grateful Dead band member Robert Hunter who co-wrote Bob Dylan's song 'Duquesne whistle', the opening track on his 2012 album *Tempest*. 'Duquesne whistle' is a catchy song with an old-time shuffle feel. It chugs along with a repeated whistle motif and train rhythms played on guitars (steel, electric and acoustic) and piano. But the jaunty music is juxtaposed with apocalyptic visions in its lyrics, where the blowing of the Duquesne train whistle is going to sweep the world away and blow the sky apart. Dylan's first recording of a train song was in 1962 when he included 'Freight train blues', replete with train whistle imitations, on his debut album *Bob Dylan*. The first version of 'Freight train blues' had been written nearly 30 years earlier by John Lair and recorded by Roy Acuff (see page 227). Dylan copyrighted his arrangement in 1978, but the two versions have much in common.[29] The 1965 song 'I'll keep it with mine' has a couple of lines about trains. 'Train A-Travelling' is a 1968 protest song where the train is used to symbolize the politics of society with its 'firebox of hatred' and 'furnace full of tears'. References to trains in other Dylan songs are less substantial. 'Slow train', a track on Dylan's 1979 album *Slow Train Coming*, which followed his conversion to Christianity, has few references to trains other than the repeated line telling us that a slow train is coming round the bend. Similarly, despite its enigmatic title, 'It takes a lot to laugh it takes a train to cry' includes only a couple of references to trains.

Three train songs for children

I'm a train by Albert Hammond

Albert Hammond and Mike Hazlewood wrote 'I'm a train' in the 1960s and Hammond's 1974 recording was the first to hit the charts. With its strumming guitars, clicking fingers and heavy use of percussion, it is strong on train imitations. It is a jolly song with straightforward lines such as 'Look at me' and 'I'm a train', but this cheeriness belies the underlying sadness of the lyrics when we are invited to look at the train for the 'very last time' and we learn where the train is heading for – the breaker's yard. The words may appear simple, but at the same time they are onomatopoeic tongue twisters, particularly in the repeated phrase 'I'm a chucka train'.

'I'm a train' lends itself to arrangement, and two of the most successful arrangements have been performed by the New Seekers and the Kings Singers. The original song uses simple chord progressions, but Peter Knight's a cappella arrangement for the King's Singers is rich in chromatic harmony coupled with intricate overlapping rhythms. There is much play on the words 'chooka chooka' 'unpitched and unvoiced' in the opening, perhaps an early example of beat boxing.

The runaway train as sung by Vernon Dalhart

'The runaway train' was written nearly 100 years ago by Robert E Massey, Harry Warren and Carson Robison and was first recorded in 1925 by the American country singer Vernon Dalhart accompanied by engine sounds, bells and whistles. The lyrics recount that the runaway accident happened in 1889 on the 'old Chicago line' when the 'rails were froze' and the air brakes of the No. 9 train wouldn't hold. During the 1880s, railroad accidents killed hundreds of crew members and passengers in the US. Brake failure was a common cause (see pages 231-232). In common with many other train disaster songs, the parts played by the crew are the focus of the lyrics; 'The runaway train' describes the actions of the engineer, the fireman, the porter, and the conductor in separate verses, as they try in vain to save the train. The most familiar lines are found in the chorus when the train comes down the track with 'whistle wide' and 'throttle back' and the repeated words 'she blew'. The song became popular in the UK when it was recorded by Michael Holliday in 1956 and a few years later by the television puppet pigs, Pinky and Perky.

Morningtown ride by The Seekers

Written by Malvina Reynolds and a hit when recorded by The Seekers in 1967, 'Morningtown ride' uses a train journey as a metaphor for sleeping. We hear the train whistle blowing as it sets out on its way, 'rocking rolling riding' on its night time journey to the sunshine of Morningtown. Judith Durham, the lead singer of this Australian folk-oriented group, is noted for her purity of tone and the sentiments of the song, essentially a lullaby, are well-suited to her mellifluous voice. The simplicity of the melodic line has led to its popularity for singing in schools.

Train journeys from around the world

Many of the songs in this chapter are set in the USA or the UK, but in this section a disparate selection of songs covers other parts of the world - Morocco, Jamaica, Germany, Cuba, Canada and South Africa. It opens with a celebration of the hippie trail to Marrakesh, moving on to a Jamaican reggae song about the spiritual journey to heaven, then a trip across Germany with the synthesised sounds of Kraftwerk, beatboxing in Cuba, the first transcontinental route across Canada and finally a heart-breaking jazz number, a tribute to the thousands of conscripted migrant labourers who were forced to work in the gold mines of South Africa.

Morocco

Marrakesh Express by Crosby, Stills and Nash

Railways in Morocco were introduced by the French Protectorate in the 1920s, and in 1923 the first railway station was constructed in Marrakesh. By the 1960s, Marrakesh was fast becoming a popular stop-off point on the Moroccan hippie trail. Graham Nash's inspiration for the song was his 1966 venture along the route that took him by train from Casablanca to Marrakesh.

Three years later he left his Manchester band The Hollies to join David Crosby and Stephen Stills in what was to become one of the first American supergroups, some would argue the voice of the Woodstock generation, anti-Vietnam, anti-social injustice, pro-hashish and spiritual enlightenment. The lyrics to 'Marrakesh Express' suggest an alternative lifestyle which looks beyond Western capitalism, having to get away 'to see what we could find', and 'sweeping cobwebs' from the mind. The song is catchy and upbeat, most of its vaguely Eastern feel comes from the lyrics. Nash describes what he could see: ducks, pigs and chickens, an 'animal carpet wall-to-wall' and 'charming cobras in the square'; exotic colourful clothing; along with a whiff of psychedelia, 'blowing smoke rings from the corners of my mouth'.

Jamaica

Roots train by Junior Murvin

'Roots train' is the opening track on Junior Murvin's 1977 debut reggae album *Police & Thieves*, which was produced in Jamaica by Lee 'Scratch' Perry. Opening with a train whistle sound, clickety-clack train rhythms on the drum kit introduce reggae roots 'train number one'. Following in the line of many spirituals and gospel music songs, it is essentially a song about spiritual salvation. It uses the railroad trip as a metaphor for the path through life where if you want to get on board then 'you gotta be righteous' and clean in 'thought, word and deed' to guarantee your place in heaven, a land where 'everything is great'.

Germany

Trans-Europe Express by Kraftwerk

'Trans-Europe Express' is the title track of the German band Kraftwerk's 1977 album of the same name. Making innovative use of early synthesiser technology to capture the sounds of a train, Ralf Hütter and Florian Schneider evoke a railway journey by means of a relentless rhythm. The lyrics use a repeated refrain of 'Trans-Europe Express' interpolated with place-name announcements – Paris, Vienna, Dusseldorf – as they go from 'station to station'. 'Station to station' is a deliberate reference to Bowie's album. Hütter and Schneider had met David Bowie prior to the recording, and this meeting is also referenced in the song's lyrics.

Cuba

El Tren by Vocal Sampling

In this 1997 recording of Rafael Cueto's song, the six members of the Cuban *a cappella* vocal group Vocal Sampling accompany their singing by imitating instrumental sounds, such as flute, trumpet, and percussion using only their voices and hands. They are masters of the art of beatboxing, imitating the sounds of percussion instruments using their mouths, lips, tongues and voices. The song, sung in Spanish, is in a traditional Cuban style. The words tell of a fateful train journey; the passengers are welcomed aboard, the journey begins, but then the power begins to fail, the train has run out of fuel and the power is failing. The song is overlaid throughout by vocal imitations of trains, sirens, clicking rails and hissing steam.

Canada

Canadian railroad trilogy by Gordon Lightfoot

Singer-songwriter Gordon Lightfoot was commissioned to write this song by the Canadian Broadcasting Corporation to celebrate the Canadian Centennial in 1967. Strictly speaking it is not a song describing a journey; rather it focuses on the building of the transcontinental Canadian Pacific Railway which was completed in 1886. However, the 'iron road runnin' from the sea to the sea does take us through the 'wild majestic mountains' the dark forests, the Rockies and the 'wide prairies' painting a picture of the varied geography of that vast country. With its simple repeated chord progressions, strumming guitars (12-string, acoustic and bass) and harmonica, 'Canadian railroad trilogy' could best be described as folk-pop. Strumming guitars set the railroad chugging along with the speed picking up as the song gets underway. Fast-paced sections at the opening and close of the song are set around a poignant slow middle section describing the isolation and hardship of the navvies.

Canada's first transcontinental railway line is 2700 miles long running from Ontario to the Pacific Ocean and it covers much difficult terrain. Traversing two mountain ranges, the Rockies and the Selkirks, it involved extensive blasting through hard rock. Around 15,000 manual labourers were employed as navvies, half of them Chinese.[30] Lightfoot's lyrics describe their long days of back-breaking work in the 'bright blazing sun', ending with the words 'many are the dead men/too silent to be real.' In all at least 800 men died. At the same time as chronicling the hardships, Lightfoot recognises the optimism of that period of Canada's history when people 'came from all around', setting up new industries and looking to the future.

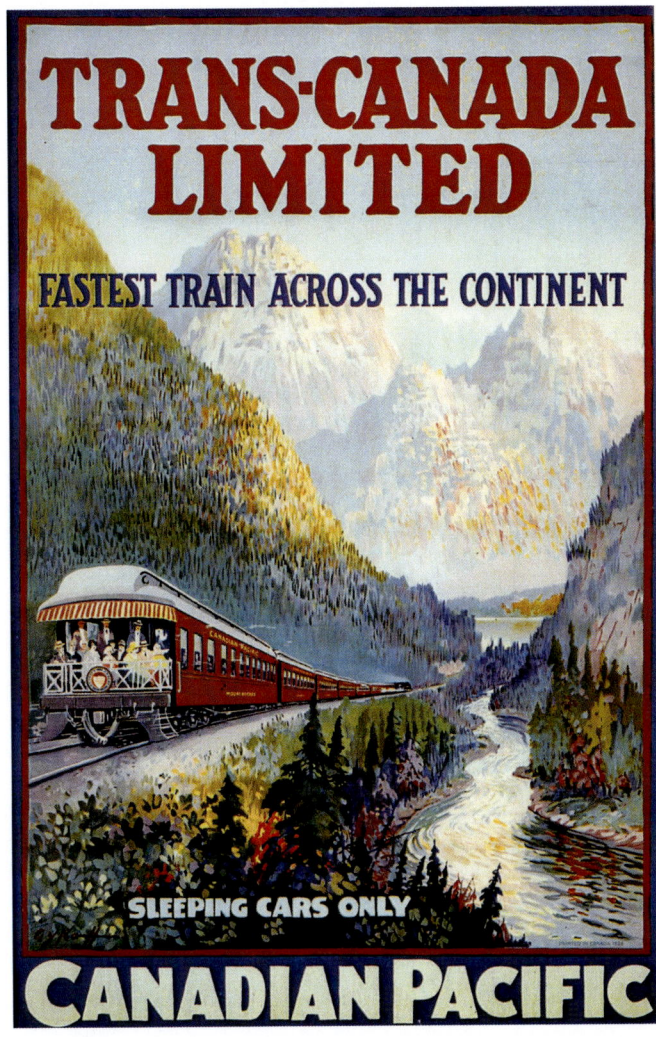

Canadian Pacific poster
Granger Historical Picture Archive / Alamy Stock Photo

South Africa

Coal train (Stimela) by Hugh Masekela

The South African jazz musician Hugh Masekela (1939 – 2018) was brought up in the mining town of Witbank in the Eastern Transvaal, a town with a population which included many migrant workers from Mozambique. For most of the twentieth century the gold-mining industry of South Africa relied on black migrant labour. The gold mines were some distance from thickly populated areas that could supply the vast low-waged workforce needed to make deep-level gold mining a profitable proposition. This meant that 'rural recruits were obliged to shuttle to and fro …on nine-month labour contracts'.[31] Between 1911 and 1930 an annual average of about 700,000 were railed to the mines. The train facilities were basic, and the treatment the migrant passengers received 'reflected the abuse, disinterest and intolerance that affected all African passengers.'[32] The historian Charles van Onselen provides a grim account of the trains that transported migrant workers from Mozambique to and from the Rand mines – journeys that he describes as 'mobile incarceration'. As late as the 1920s some workers still had to travel part of the 20 hour journey in open coal trucks during the cold winter with little food or water.[33] The Congolese author Fiston Mwanza Mujila wrote that trains 'had a whole other symbolism than in Europe. They symbolized the taming of African nature, deportation, forced labor, exploitation, the transport of minerals, looting etc.'[34]

When Masekela was at school he was famously given his first trumpet by Archbishop Trevor Huddleston, a champion of anti-apartheid and the chaplain at Masekela's school. As well as playing trumpet and flugelhorn, Masekela was a powerful singer and songwriter with a strong political voice. He became an active anti-apartheid supporter, which eventually led to his exile from South Africa.[35]

He recorded several versions of this powerful song. It opens with repeated beats on the cow bell, a simple but effective way of evoking the sound of a moving train. This leads to a spoken introduction in which Masekela lists areas of Africa that the conscripted labourers had been transported from to work in the 'golden mineral mines of Johannesburg'. He describes the cruel conditions in which they worked, the 'stinking, funky, filthy, flea-ridden barracks' where they lived and how they were cut off from their families who may have been 'forcibly removed' or 'wantonly murdered' in their absence. To the migrant workers the

'Choo-choo', as Masekela sarcastically refers to the coal train, is not an object of affection; rather they 'curse it'. The spoken section ends with a dramatic 'Whooa' reminiscent of a steam train whistle, and the cow bell leads into Masekela's horn solos (either flugelhorn or trumpet on different recordings). 'Coal train' is a soulful song 'blistering and mournful, infused with the energies of resistant people', as Sharae Deckard puts it 'Hugh Masekela wields trumpet and words like weapons, introducing a deeply political consciousness of class struggle and social justice into his fusion of AfroBeat and jazz funk'.[36]

Into the twenty-first century

In 2001 Pennsylvanian singer-songwriters, Artese 'N Toad, released the album *They Don't Write Songs About Trains Anymore*. The acoustic Americana songs are all centred on train themes, stories about railroad events and people, with plenty of historical detail. The album proved to be popular, partly because it was given away with model train sets made by MTH Electric Trains, and its popularity led to a follow-up album. The album *Traineater* by the Brooklyn-based rock band Book of Knots mourns the decline of heavy industry in the 'Rust Belt' ' and its title song, 'Traineater', is about 'Old No. 6, a locomotive on its way to be scrapped. It also includes a version of 'The ballad of John Henry'. Similarly the English progressive rock band Big Big Train have released two albums including songs with historical train themes: *The Underfall Yard* and *English Electric Vol. 2* where 'East coast racer' tells the story of the record breaking run by *Mallard* over 80 years ago.

All of the above are songs of nostalgia for a bygone age, performed in a musical style which is no longer at the forefront of contemporary popular music. So is it true that they don't write songs about trains anymore? Are there any twenty-first-century pop songs celebrating current rail travel, and are any of them in one of the styles which currently predominate the charts, namely electronic dance music, ambient and hip-hop, which have seen an unrivalled rise in popularity worldwide? Certainly they are few and far between in comparison with the output of train songs during the nineteenth and twentieth centuries.

However, at least one pop song has been written to celebrate a landmark in the development of the railways during this century. In 2003, the Central

Japan Railway Company commissioned Rei Nakanishi and Kyōhei Tsutsumi to write a song to celebrate the opening of a new area of Tokyo's Shinagawa Station created to accommodate the Shinkansen (bullet trains). The song 'Ambitious Japan', performed by the band Tokio, was used as part of a huge campaign and the Bullets' 300 and 700-series trains were emblazoned with its title; it soon reached No. 1 in the Japanese charts. The trains may have been cutting edge, but the musical style was the type of pure pop which has been around for decades.

There are some references to trains by American rock bands in the early 2000s. In their song 'Clark Gable', indie band The Postal Service open with a familiar theme when they sing that they are waiting for a train in the London Underground. Julian Casablancas, the lead singer of rock band The Strokes, compares himself to a train 'moving too fast' in 'Automatic stop' and in the sinister chorus of 'Enfilade', hard-core rock band At The Drive In scream of being a sacrifice on a railroad track, tied and gagged with a freight train coming. A search for railway songs in a more contemporary idiom reveals that trains have not totally lost their appeal. In 'Girl on a train' (2016), Harlem rapper Skizzy Mars tells of falling in love with a pretty stranger on the 'L with some headphones on'. In the same year hip-hop duo Macklemore and Ryan Lewis rap from the perspective of a disconnected traveller who uses the train as a means of escape. In this song the train is used in the now familiar metaphor of a journey passing through life; as we hear in the chorus 'Otra cuidad, otra vida' (Another train, another life). There are no lyrics in the lo-fi ambient instrumental 'Train ride home' (2018) by lofi.samurai, but the title represents another common theme found in train songs.

Perhaps the heyday of the train song has passed, but railways are still able to illuminate the wider aspects of life: acting as a metaphor or representing a means of escape; taking you away and bringing you back home; connecting places and people. They can be a place of reverie, of frustration or romance, and their evocative sounds and irresistible rhythms continue to inspire countless pieces of music.

Endnotes

1. T R Gourvish. *British Railways 1948-73. A Business History.* (Cambridge: Cambridge University Press, 1986): 437.
2. Charles Loft. *Government, the Railways and the Modernization of Britain: Beeching's Last Trains.* (Abingdon: Routledge, 2006).
3. In 1965 Widnes station was named Widnes North station.
4. *The Guardian*, 1 August, 1990.
5. Spencer Leigh. *Simon & Garfunkel. Together Alone.* (Carmarthen: McNidder and Grace. 2016).
6. *The Times*, 7 October, 2000.
7. Elsewhere the Windsor folk club is referred to as the Howff folk club which was housed in the Windsor Rooms in Widnes.
8. Robert Hilburn. *Paul Simon. The Life.* (London, Simon and Schuster UK (2018)): 79-80.
9. *The Guardian*, 25 April, 2001.
10. Hilburn, *Paul Simon*, 79-80.
11. Simon Bradley. *The Railways: Nation, Network and People* (London: Profile Books, 2015): 274
12. *The Times*, 21 October, 2009.
13. According to the CBC, there would not have been a midnight train to Georgia in 1973, nor for that matter a through train. The best route would be via New Orleans with the train leaving Los Angeles at 9.00 pm to make an overnight connection. However, Weatherley's song started its life as the 'Midnight Plane to Houston' referencing a comment made in a telephone conversation with his friend Farrah Fawcett. https://www.cbc.ca/radio/the180/the-problem-with-this-song-midnight-train-to-georgia
14. There is more than one recording of this song with either Bob Marley or Peter Tosh as the vocalist.
15. Kwame Dawes. *Bob Marley. Lyrical Genius.* (London: Bobcat Books, 2002).
16. https://thequietus.com/articles/03195-the-damned-s-captain-sensible-on-why-he-likes-trains
17. A flanger works by mixing the original sound with a very slightly delayed version, whereas delay effects create echoing sounds.
18. Peter Doggett. *The Man Who Sold the World: David Bowie and the 1970s.* (London: Vintage, 2012).
19. Nicholas Pegg. *The Complete David Bowie.* (London: Titan Books, 2016).
20. Panning is the technique of shifting a sound within the stereo field so that it appears to move from one speaker to another.
21. Pegg. *The Complete David Bowie*, 2016.
22. A dictionary definition of the word 'parapliers' has not been found.

23 https://tomwaitspodcast.wordpress.com/2016/03/09/i-lived-my-life-on-dreams-and-trainsmoke-tom-waits-trains/

24 https://jimjarmusch.tripod.com/snc93.html

25 Ibid

26 http://www.rattenne.com/essay/dedtrain.html

27 Gopal Das. 'Cocaine Abuse in North America: A Milestone in History'. *The Journal of Clinical Pharmacology* 33(4) (1993):296-310 .

28 Ibid.

29 For more information and the lyrics to 'Freight train blues' see pages 227 and 270.

30 Christian Wolmar. *A Short History of Trains.* (London: Dorling Kindersley Ltd., 2019): 125-127

31 'Gordon Pirie. 'Brutish bombelas. Trains for migrant gold miners in South Africa c. 1900-25' *Journal of Transport History* 18 no. 1 (1997): 31-44.

32 Ibid.

33 Charles van Onselen. *The Night Trains: Moving Mozambican Miners to and from South Africa, circa 1902-1955.* Oxford: Oxford University Press, 2019).

34 As cited in Kathy Mansfield. *Noel's story: A man of Zimbabwe.* (Kibworth Beauchamp: Matador, 2020).

35 Robin Denselow *When the music's over. The story of political pop.* (London: Faber & Faber, 1989): 48-9.

36 S. Deckard and S Shapiro (eds.). *World literature, Neoliberalism and the Culture of Discontent.* (London: Palgrave Macmillan, 2019): 246.

General Index

SONGWRITERS, ARTISTS AND their album titles will be found in the 'Popular music index' under the artist's name. Works by classical music composers will be found in the 'Index of composers and their works' under the composer's name.

Abbey, M E 204
    accidents, train 6-8, 22-4, 64, 78, 84, 104, 108, 114, 162, 188, 208, 210-12, 230-4, 271
    Europe 84, 104, 108
    UK 6-8, 22-4, 64
    USA 78, 188, 208, 210-12, 230-4, 271
Alabama 201, 208, 214, 215
Aldrich, Mark 231
ambient music 277, 278
American Library of Congress II, 190
Anthony, Scott 129
Arkansas 228, 229, 253
Auden W H 127, 129, 131, 133
Austria II, 75, 89

Ballet Russes 142
Barry, Robert 76, 78
beatboxing 272, 274
Beeching, Richard 257, 258, 265
Beerbohm, Max 33

Belgium 77, 97, 109
Belle Vue, Manchester 47-9, 52
Bennett, Arnold 33
Best Friend of Charleston 187
Bienvenüe, Fulgence 146
Birmingham Special 251
Blackmore, John 21
Bloomfield, Howard 242
blue notes 242
bluegrass 204, 240-1, 252
blues 187-94, 196-203, 210, 213, 224-27, 239-48, 252-3, 267
    delta blues 198, 199
    jump blues 253
Bologna 181, 182
boogie-woogie 239, 242, 243-6, 252
Bosworth, John 40
Bradley, Simon 7, 10, 110, 261
brass bands 27, 45-56, 58-60, n. 60, 161, 180
brass band contests 45, 47-52

Brazil and Brazilian music 137, 143, 147-50
Brief Encounter (film) 179
broadside ballads I, 3-24, 120, 167, 189, 230
Broady, Joseph Andrew "Steve" 232-3
Brown, David 119
Brunel, Isambard Kingdom 21
bullet trains 278

Calais 144, 168, 170
camp meetings 203
Canada 187, 203, 272, 274-5,
canals 7, 21, 35
Cannonball Express 211
Carlisle 6, 13, 15, 19, 21
Carnforth 179
Catherine the Great 117
Chanel, Coco 144
Chappell, Louis Watson 208
Chattanooga 251-2
Cheap Trains Act 1883 40
Chicago 202, 211, 217, 222, 227, 243, 262, 271
Cocteau, Jean 138, 144
Cohen, Norm 196-7, 204-5, 218, 230
Coleman, Terry 9-10
Copenhagen II, 77, 89-92, 95-6
country music 188, 196-7, 246-7
Cranston, Catherine 132, n. 135
Crewe 11, 34, 38, 54, 130
Cuba 256, 272, 274

Dawes, Kwame 263
Denmark 89
Detroit 187, 194, 243, 262
Diaghilev, Sergei 142, 144
Dixie Flyer 200, 202-3, 240
Dixieland 197, 200, 202, 227

Doncaster 36, 49, 54, 58
Dorson, Richard 208
D'Oyly Carte Opera Company 63
Dubois, Pierre 103, 104
Duke of Sutherland 68

Edinburgh 23, 58
electroacoustic music 173-4, 175
electronica 256, 267
electronic dance music (EDM) 174, 277,
electronic instruments 175, 266
Eliot, T S 33, 137
Ericson, Nils 89
excursion trains 29-30, 34-7, 45, 47, 50-2, 56, 85, 230, 240

Fairfax, Bryan 154
Fast Flying Virginian 230
fiddle 76, 155, 164, 193, 239, 240-1
fiddle music 193, 239, 240-1
first class carriages 63, 110
Flaubert, Gustave 79-80
  Madame Bovary 79-80
Flying Scotsman 23, 24, 49
formalism 165, 166
Fox, Christopher 178
France 97, 101, 102, 118, 145, 168-9, 174, 251
Friedman, Martin 182-3
Futurism II, 12, 139, 146

Gabler, Milt 253
galop II, 90-2, 94-
gaols (see prisons and prisoners)
Garon, Paul 194
George, David Graves 234
Germany 76-7, 97, 162, 273
Gerstner, Franz Anton von 117

GENERAL INDEX

Gibbons, Jack 98
Giles, Francis 21
Gioia, Ted 189, 191
Golden Spike 188
Goldsmith, Lyn 260
Goole 180
Gordon, Robert Winslow 190
gospel music 187, 203-5, 227, 273
GPO Film Unit 127-30
Grainger Museum 153
Grayrigg 15
Great Depression 175, 192, 243
Great Exhibition, The 14, 35, 50
Greece 75-6
Griffiths, Paul 179
Gripp, Parry 264
Groves, George 50
Guimard, Hector 146

hammer songs 189
Hardy, Thomas 163
Harlem 247, 250, 278
harmonica 239-40, 241, 265, 268, 275
Hawthorne, Nathaniel 156, 159
Celestial Railroad, The 159
Henry, John 191, 207-10
Hilburn, Robert 260-1
hillbilly music 197, 209, 234
hip-hop 175, 277, 278
HM Railway Inspectorate 22
hoboes II, 175-6, 188, 192-6, 197, 198, 199, 205, 221-2, 231
Hobsbawm, Eric 73
Holman, Gavin 53
Holmfirth 21-3
Holocaust, the 1, 178-9
Huddersfield 21-2
Huddleston, Trevor 276
Hull 5, 35, 50, 180

Huntley, John 28-9
hymns 156, 157-9, 160, 179, 204

Industrial Revolution 56, 73, 242
Italy 153-4, 170-1, 181

Jackson, Enderby 50-2
Jamaica 208, 263, 272, 273
James, Frank 216-7
James, Jesse 207, 216-8
Janin, Jules 103, 105, 106
jazz 42, 129, 138, 143, 162, 169-70, 187, 202, 239, 242, 247-52
Jenny Lind 46
Jewell, Derek 247-8
Johnson, Julian 81
Jones, Casey 207, 210-13, 255, 269-70

Kemp, Peter 78
Kendal Fair 6, 14
Kipling, Rudyard 28
Kugler, George J. 228
Kukolnik, Nestor 118

Laube, Heinrich 80
Leach, MacEdward 208
Leno, Dan 29, 31, 41
Le Train Bleu 144
Liebhold, Peter 188
Lille 97-8, 103, 104, 168
Liverpool 3, 6, 11, 18, 34, 40, 46, 73, 166, 260
Lomax, Alan II, 190, 204, 222, 223-4, 229, 242-3
Lomax, John A II, 190, 204, 218, 224, 228, 229
London 27, 30, 33-4, 42
    Crystal Palace 50-5
    London Underground 30, 278

London Olympics 169
Victoria Station 155, 171
London railway lines 15, 127, 132-3, 170-1
London railway companies 15, 16-17, 34, 58, 68

Mallard 49, 150, 277
Manchester 151, 153
Mansfield, Kathy 277
marches 49, 91, 98, 113, 155, 156
Marshak, Samuil 165
Maxwell, Glynn 169
Melbourne 151, 153
Mellers, Wilfrid 155
Mellotron 266, 267
Midland Railway 34, 40, 55
Milburn, George 221-2
    Hobo's Hornbook, The
Mississippi 191, 194, 196, 198, 223, 227
Mitry, Jean 140
Morocco 272-3
Morrison, Blake 180
Morrison, Richard 261
Mozambique 276
Mujila, Fiston Mwanza 276
Murray, Albert 194, 247-8
music hall III, 5, 27-42
Music Hall Artistes Railway Association, The (MHARA) 39-42
music publishing and printing 4, 24, 27-8, 77 82, 133-4, 249
musique concrete 174, 266

Napoleon III 97
National Union of Railwaymen (NUR)

navvies 3-12, 274
Nelson, Scott Reynolds 189
New York 78, 134, 140, 141, 152, 157-9, 164, 177, 187, 218, 241, 247, 251-3
North Carolina 190, 214, 223, 226, 232
Norway 89, 90
nostalgia II, 257, 269, 277

O'Connell, Daniel 12, n. 25
Oliver, Paul 193, 201, 202, 221,
Onselen, Charles van 276
Orient Express, The 161, 170-2
Osborne, Richard 111

Pace, Kelly 228, 229
Pacific 231 (film)
Paris 46, 77, 92, 102-4, 107, 110-12, 137-8, 141-2, 145-47, 174, 273
    railway accidents 101, n. 108
    railway lines 97-99, 103, 118-19, 144, 167, 169, 170-1
Paris Conservatoire 137, 138, 145
Paris Metro 97, 145-6
Parliamentary trains 64-5
Paton, Alan 164
Picasso, Pablo 137-8, 144
Piccadilly Station, Manchester 179
Polenberg, Richard 207-8
Polka 49, 79, 83-6, 91
Ponce de Leon 231
porters 28, 29, 31-4, 37-9, 178, 258
Power, Dominic 173, 179
Priestley, J B 128
prisons and prisoners 189, 190, 193, 208, 222-4, 228-9
psychedelia 266, 273

GENERAL INDEX

Railroad Bill 207, 214-16,
railway choirs 45, 56-8
Railway Clearing House 41, 56
Railway mania 4, 6, 7, 46
railway musical societies 56-8
railway line openings I-II, 80, 161, 167
    United Kingdom I-II, 3, 4, 6, 18-22, 34, 45, 169
    Europe II, 89-90, 97-8, 103, 168-9, 170
    Russia 117, 120
    Japan 277-8
Railway Regulation Acts 22, 36, 40, 64-65
railway workshops 45, 46
Rattenne, Ken 269
rent parties 243
Revill, George 139, 149,
rhythm and blues (R&B) 239, 252, 253
Rio de Janeiro 143, 147, 148
Rolt, L T C 22
Romain, Jules 145
Rosen, Adolf Eugene von 89
Royal Palm 231
Rubinstein, Artur 144
Russia 77, 83, 86, 117-20, 165, 167

St. Petersburg 77, 83, 90, 92, 117, 118-20
    Pavlosk 77, 91, 95, 117
Salford 166
samples and sampling I, 173-5, 176, 178, 266
Sandburg, Carl 12, 213, 215, 224
Scandinavia 89-91
Schafer, R Murray 241-2
Schrökl, Gustav and Therese 85
Scowcroft, Philip n. 69, n. 150

second class carriages 63, 64
Sharp Brothers and Company 89-90
sheet music 27, 28, 213
Shillibeer's Original Omnibus 4, 6, 16-17
skiffle 210, 229
Slater, Morris (see Railroad Bill)
slaves and slavery 188, 192, 203, 214
Smith, Richard 48
smoking compartments 36
smoking concerts 57
Socialist Realism 165
Solovox 153, n. 160
South Africa III, 256, 272, 276
Southern Pacific Golden Gate Limited
Southern States II, 190, 196, 200, 205, 239
Spector, Phil 257
spirituals II, 203-5, 273
Sportsman 231
Stannard, Robert 166-7
Stephenson, George 3, 21, 109, 166
Street, John 28
string bands 240-1
Sun Valley Serenade 251
Sweden 89, 90

Tamla Motown 262
Texas 197, 224, 243, 246, 247
Theremin 153, 157, n. 160
third class carriages 63, 64, 110, 163, 170
Tivoli Gardens (Copenhagen) 77, 91
Tokyo 171, 278
Tolstoy, Leo 167
    Anna Karenina 167
Tom Thumb 187
track gauge 75, 89, 97, 117

train bells 65, 84, 92, 93, 112, 135, 146,
    178, 202-3, 242, 248-9, 269,
    271
train whistles I, 65, 83, 84, 91-4, 179,
    211, 221-2, 227-8, 240-2, 270,
    271
Tsar Nicholas I 117
Turksib 129, n. 135

Underground Railroad 203, 205

Vaurabourg, Andrée 139
Venice 170-1
Vienna 75, 76-8, 80, 84-5, 92, 162, 273
    Zum Sperlbauer (the Sperl) 76, 80
Vietnam war 264-5, 273

Wabash Cannonball 221-2
Wade, Stephen 223, 228,
waltz II, 49, 76, 77-83, 90, 91, 120, 134
Weimar Republic 162
Weissman, Dick 198
Widnes 255, 260-1, n. 279
Wigan 261
Wolfe, Charles 241, 254
Wolmar, Christian 117
Wordsworth, William 13, 16
work songs II, III, 1, 187, 188-91, 205,
    228
World War I 7, 42, 58, 145, 152
World War II 7, 42, 52, 58, 73, 134,
    145, 178, 251, 253

York 46, 58, 59

Zola, Emile 167

Classical Music Index

Alkan, Charles-Valentin 97, 98-102, n. 108
   Funeral March on the Death of a Parrot 98
   Le Chemin de Fer (The Railway) 98-102, n. 108
   Les omnibus 102
   Twelve Studies In All The Minor Keys, Opus 39 98
Auric, Georges 138

Bach J S 138, 147, 148
Bedford, David 266
Bellini, Vincenzo 51, 118
Berio, Luciano 176
Berlioz, Hector 77, 97-8, 102-6, n. 108, 118
   Grande symphonie funèbre et triomphale 104
   La damnation de Faust 103
   Le Chant des Chemins de Fer (The Song of the Railways) 103-6, n. 108
   Symphonie Fantastique 102
Boulanger, Nadia 137
Boulez, Pierre 173

Britten, Benjamin III, 125, 127-33, n. 135, 163
   Coal Face 131
   Night Mail III, 127-33, n. 135
   Way to the Sea, The 125, 127, 132-3
   Winter Words 163
Bryars, Gavin 172, 179
   The Stopping Train 172, 179

Cage, John 172, 180-82, n. 183
   4' 33" 180
   Alla Ricerca del Silenzio Perduto (Il Treno) 172, 180-82,
Candish, C F Chudleigh 58
   Song of the Jolly Roger 58
Chopin, Frederic 98, 113, n. 115
Copland, Aaron 137, 139, n. 150, 210
   John Henry 210

Davies, Peter Maxwell 165, 166-7
   Chat Moss 166-7
   Eight Songs for a Mad King 166
   Symphony No. 5 166
Debussy, Claude 98, 138
Dhomont, Francis 175
   Espace 175
Durey, Louis 138

Ellis, Vivian 125, 133-5
   Bless the Bride 134
   Coronation Scot 125, 133-5
   Mr Cinders 134
   Spread a little happiness 134
   This is my lovely day 134

Finnissy, Michael 169-70
Freightrain Bruise 169-70
Fitkin, Graham 161, 169
   Track to Track 169
Foulds, John 129
Fučík, Julius 94
   Entry of the Gladiators 94

Gershwin, George 146, 153
   American in Paris 146
Gilbert and Sullivan 63-8, n. 69, 119, 134
   HMS Pinafore 63
   Judge's Song, The (from *Trial by Jury*) 64
   Junction Song, The (from *Thespis*) 65-8, n. 69
   Lord Chancellor's Nightmare Song, The (from *Iolanthe*) 63-64
   Railway Song (see The Junction Song)
   Mikado, The 63, 64
   Pirates of Penzance, The 63
Glass, Philip 177
Glinka, Mikhail 117-20
   A Life for the Tsar 118
   Ruslan and Lyudmila 118
   'The Travelling Song' from *A Farewell to St Petersburg* 117, 119-20

Grahl, Traugott 90
   Sveas helsning till Nore, Walzer (Greetings from Sweden to Norway, Waltz) 90
Grainger, Percy 151-6, 160, n.160
   Arrival Platform Humlet 154-5, 160
   Country Garden 151
   Free Music no. 2 153
   Lincolnshire Posy 153
   To a Nordic Princess 152
   Train Music 151, 153-4, 160
Gung'l, Josef 94-6
   Eisenbahn-Dampf-Galopp (Steam Railway Galop) 94-6

Handel, G. F. 45, 48, 51, 56
   Messiah 56
   See the conquering hero comes 45, 48
Henze, Hans Werner 161, 164
   Boulevard Solitude 164
Honegger, Arthur II, 137-42, 149, n. 150
   L'Aiglon 138
   Pacific 231 II, 137, 139-42, 146
Hoyer, Franz 90-1
   Jernban-Galopp (Railway Galop) 90-1

Ibert, Jacques 137, 138, 145-6, 149
   L'Aiglon 138
   'Le Metro' from *Suite Symphonique* 137, 145-6

Ives, Charles 151, 156-60, 179
   Celestial Railroad, The 151, 159-60, 178
   Concord Sonata 156
   'From Hanover Square North' from *Orchestral Set No. 2* 151, 157-9, 179
   Three Places in New England 156
   Unanswered Question, The 156

Krenek, Ernst 161-3
   Ballad of the Railroads, The 162
   Jonny spielt auf 162-3
   Santa Fe Timetable 162-3

Lanner, Joseph 76, 91
Les Six 137, 138, 143
Lindblad, Otto 90
   Malmö Järnbanesång (Malmö Railway song) 90
Lumbye, Hans Christian 11, 90, 91-6
   Copenhagen Steam Railway Galop 11, 90, 92-6
   Tivoli Bazaar Galop 92
   Tivoli Shooting Gallery Galop 91
   Tivoli Steam Merry-go-round Galop 92

Marchant, Stanley 58
Meyer, Jean 90, 91
   Jernvägs-Galopp (Railway Galop) 90
   Milhaud, Darius 137-8, 143-4, 147, 177
   Le Boeuf sur le toit 143
   Le Train Bleu 137, 144, 149

Nyman, Michael 161, 168
   MGV (Musique à Grand Vitesse) 168

Partch, Harry 173, 175-6
   Genesis of a Music 176
   U.S. Highball 173, 175-6
Patterson, Paul 161, 168-9
   Royal Eurostar 168-9
Poulenc, Francis 137, 138, 142-3, 149
   'En chemin de fer' from *Promenades* 137, 142-3
   Gloria 142
   Stabat Mater 142
Prévost, Abbé 164
   Manon Lescaut 164
Prokofiev, Sergei 147, 161, 165-6
   Peter and the Wolf 165
   Romeo and Juliet 165-6
   Winter Bonfire 165-6

Reich, Steve 1, 173, 176, 177-9
   Clapping Music 177
   Come Out 177
   Different Trains 1, 173, 176, 177-9
   It's Gonna Rain 177
Riley, Terry 177
Rossini, Gioachino 51, 98, 109-14
   Péchés de vieillesse (Sins of Old Age) 111
   Un petit train de plaisir comico-imitatif (A little train of pleasure) 109, 111-14

Satie, Erik 111, 138
Schaeffer, Pierre 173-5
   'Etude aux chemins de fer' from *Cinq études de bruits* (Five Studies of Noises) 173-5

Schubert, Franz 129
   Rosamunde 129
Seeger, Ruth Crawford 226
Sparke, Philip 161, 170-2
   Orient Express 161, 170-2
Strauss, Eduard 76, 78, 79, 83-6
   Bahn Frei! (Line Clear) *Op.* 45 79, 85-6
   Feuerfunken (Sparks of Fire) *Op.* 185 79, 82
   Lustfahrten (Pleasure Journeys) *Op.* 177 79, 85
   Mit Dampf (With Steam) 79, 84
   Ohne Aufenthalt (Non Stop) 82, 84
   Ohne Bremse (Without Brakes) 82, 84
   Tour und Retour (Round Trip) 79, 85
Strauss, Johann I 76-7, 79, 80-2
   Eisenbahn-Lust-Walzer (Railway Pleasure Waltzes) Op. 89 80-2
Strauss, Johann II 76, 77, 79, 82-3, 85, 86
   Accellerationen (Accelerations) Op. 234 79, 82-3
   Reise Abenteuer (Travel Adventures) 86
Spiralen (Spirals) Op. 209 79, 83
   Vergnügungszug (Pleasure Train Polka) Op. 281 79, 83, 85
Strauss, Josef 77-8, 79, 85
   Gruss an München (Greetings to Munich) Op. 90 79, 85
Stravinsky, Igor 138, 147, 154
   Rite of Spring 154
Sullivan, Arthur 63, 65, 69, 119, 134

Villa-Lobos, Heitor 137, 147-9
   'The Little Train of the Caipira' from *Bachianas brasileiras No. 2* I, 137, 148-9

Wagner, Richard 111, 138
Weill, Kurt 161, 164, n. 172
   Ballad of Mack the Knife, The 164
   Lost in the Stars 164
   Railroads on Parade 164
Wiegold, Peter 172, 179
   The End of the Line (A Brief Encounter) 173, 179-80
Wolfe, Julia 210
   Steel Hammer 210
Woodgate, Leslie 58

Xenakis, Iannis 174

Popular Music Index

SONGS ARE LISTED in the 'Index of song titles'.

ABBA 224
Acuff, Roy 192, 222, 227, 270
American Music Shop Band 241
Artese 'N Toad 277
    They Don't Write Songs About
        Trains Anymore 277
At The Drive In 278
Autry, Gene 192, 197
Ayer, Nat D 30
Ayers, Kevin 255
    Joy of a Toy 255

Bacharach, Burt 258
Backsliders, The 264
Baez, Joan 216
Bailey, Buster 202
Bailey, De Ford 240
Ball, Bentley 218
Barber, Chris 229
Bard, Wilkie 29, 30
Basie, Count 252
Beatles, the 229, 264,
Bedford, David 266
Berlin, Irving 252

Bigard, Barney 248
Big Big Train 277
    English Electric Vol. 2 277
    The Underfall Yard 277
Big Bill Broonzy 201, 224
Bilbro, Bert Hunter 240
Blind Willie McTell 246
Blink-182 264
Block, Martin 262
Bogan, Lucille 200-1
Boggs, Dock 192
Bon Jovi 259, 264
Book of Knots 277
    Traineater 277
Bowie, David 255, 267, 273-4
    Station to Station 267
Boxcar Willie 197
Boyce, Tommy 264
Boyd, Eva (see Little Eva)
Bragg, Billy 262
Brisebois, Danielle 262
Bryden, Beryl 229
Burke, Solomon 257
Burnett, Dick 192

Captain Beefheart and his Magic Band 255, 267-9
Carlisle, Cliff 192, 197
Carr, Sister Wynona 262
Carson, Fiddlin' John 209-10
Carter Family, The 192, 205
Carter, Wilf 197
Casablancas, Julian 278
Cash, Johnny 205, 210, 230, 257, 262
Cave, Nick 267
Checker, Chubby 256
Chicago 250
Clapton, Eric 200
Clark, Guy 262
Clark, Henri 29, 38
Clarkson, Kelly 262
Clifton, Harry 29
Copeland, Martha 192
Cotten, Elizabeth 225-6
　Folksongs and Instrumentals with Guitar 226
Creedence Clearwater Revival 224
Crosby, David 273
Crosby, Stills and Nash 272-3
Cross, Hugh 192
Cueto, Rafael 274
Cutler, Scott 262
Cutrell, Dave "Pistol Pete" 224

Dalhart, Vernon 234, 271
Damned, The III, 265
Dan Hicks and His Hot Licks 262
Darin, Bobby 230
Davies, Ray 265
Dead Moon 264
Delmore Brothers, The 257
Donegan, Lonnie 210, 216, 229-30
Dorsey, Tommy 252
Durham, Judith 272

Dylan, Bob 216, 227, 255-6, 258, 269, 270
　Bob Dylan 227, 270
　Slow Train Coming 270
　Tempest 256, 270

Earle, Fred 30
Electric Light Orchestra 264
Ellington, Duke I, III, 153, 239, 247-50, 252
Elliott, Ramblin' Jack 216

Fitzgerald, Ella 250
Flanders and Swann 1, 257, 258
Fleetwood Mac 259
Formby, George (Senior) 29
Fox, G D 38
Froese, Edgar 267
　Epsilon in Malaysian Pale 267
Fyffe, Will 29

Garcia, Jerry 213, 269
Garland, Judy 252
Gates, David 264
Gilkyson, Terry 257
Gladys Knight and the Pips 262
Glazer, Joe 192, 252
Goffin, Gerry 256
Graham Central Station 264
Grand Funk Railroad 256
Grateful Dead 213, 255, 269-70
Green, Charlie 202
Green, Peter 259
Griffith, Nanci 262
Guthrie, Arlo 192, 263, 264
Guthrie, Woody 204, 205, 210, 218, 230

Hamilton, Wyzee 193

Hammond, Albert 271
Harrington, J P 29, 30, 35
Harrison, Denham 29
Hart, Bob 264, 265
Hays, William Shakespeare 204
Hazlewood, Mike 271
Henderson, Joe 250
Hennessy, Carly 262
Hidalgo, Juan 181
Holliday, Michael 272
Hooker, John Lee 192, 194, n. 206, 226
Hoots & Hellmouth 259
Hoots, Sean 259
House, Son 199
Howell, Peg Leg 192
Howlin' Wolf 265
Hunter, Harry 29, 38
Hunter, Robert 270
Hütter, Ralf 273
Hylton, Jack 252

Ives, Burl 195

James, Jesse 213
Jarrett, Michael 269
Jerry Garcia Band 262
Johnson, Robert 198-200
Jones, Rickie Lee 258
Jordan, Louis 239, 252, 253

Kent, Cissie 30
King, Carole 256
Kingston Trio, The 257
Kinks, the II, 264, 265
   Village Green Preservation Society 265
Knight, Gladys 262
Kottke, Leo 264

Kraftwerk 256, 267, 272, 273-4
   Autobahn 267
   Trans-Europe Express 267, 273-4

Lair, John 227, 270
Langtry, Lillie 29, 35
Larkin, Patty 259
Lashwood, George 30
Leadbelly II, 190, 210, 224, 228-9
Le Brunn, George and Thomas 31, 33
Ledbetter, Huddie (see Leadbelly)
Lee, Peggy 262
Leno Dan 29, 31, 41, n. 43
Lewis, Meade Lux 244
Lewis, Walter "Furry" 213, 246
Little Eva 256-7
Little Richard 230
Lloyd, Marie 29, 31-3
lofi.samurai 259, 278
Loggins, Dave 257
Longshaw, Fred 202
Lostprophets 264

McAbee, Palmer 240
McClintock, Harry 195, 218
McGhee, Brownie 216, 224
McGinty's Oklahoma Cow Boy Band 224
Macklemore and Ryan Lewis 278
Marchetti, Walter 181
Marley, Bob 262, 263, n. 279
Masekela, Hugh III, 255, 276-7
Massey, Robert E 271
Mayall, John 258
Mayer, John 264
Mayo, Sam 30
Memphis Minnie 200, 201-2
Merman, Ethel 252
Metheny, Pat 259

Miller, Glenn 239, 251-2
Mingus, Charlie 250
Minogue, Kylie 256
Mississippi John Hurt 191, 210, 226, 246
Monkees, The 264-5
Moore, Rich 259
Morris, Walter 192
Morrison, Van 218, 224, 257
Move, The 262
Muldaur, Maria 262
Mumford and Sons 204
Murray, Billy 213
Murvin, Junior 273
   Police & Thieves 273

Nakanishi, Rei 278
Nash, Graham 272
Nelson, Willie 257, 262
Newton, Eddie 213

O'Brien, Mollie 259
O'Connor, Mark 241
O'Neal, Waldo 192, 196
Oasis 267

Party Animals 262
Perry, Lee "Scratch" 273
Peter, Paul and Mary 204
Pinky and Perky 272
Pleasants, Jack 30
Pogues, The 218
Postal Service, The 278
Presley, Elvis 259
Prima, Louis 262
Primal Scream 264
Prince Buster 262
Puckett, Riley 216, 218

Rathburn, Eldon 154
Ratledge, Mike 266
Red Hot Chilli Peppers 267
Reed, Jerry 264
Reeves, Goebel 192
Rein, Trine 262
Restless Heart 257
Reynolds, Malvina 272
Rivers, Johnny 224
Robeson, Paul 210
Robey, George 29, 30
Robison, Carson 192, 271
Roche, Betty 250
Rock Island Colored Booster Quartet 227-8
Rodgers, Jimmie 192, 196-7, 262
Rolling Stones 200, 229
Rouse, Ervin T 241
Rush, Tom 262
Rutherford, Leonard 192

Saunders, Wallace 210, 212
Schneider, Florian 273
Seeger, Charles 226
Seeger, Mike 226
Seeger, Peggy 226
Seeger, Pete 210, 213, 218, 230
Seekers, The II, 272
Serendipity Singers 257
Shaver, Billy Joe 257
Siebert, T. Lawrence 213
   Talking Union 213
Simon and Garfunkel 259
   Parsley, Sage, Rosemary and Thyme 259
Simon, Paul 255, 258, 259-61
Simpson, Martin 262
Skizzy Mars 278
Skylar, Sunny 262

Smith, Bessie 200, 202-3
Smither, Chris 259
Snow, Hank 192, 197
Soft Machine 266
Spanish Town Skabeats 262, 263
Springsteen, Bruce 210, 218
Stanley, Roba and Bob 216
Stewart, Rod 269
Stills, Stephen 273
Strayhorn, Billy 247, 249-50
Strickland, Leon 246
Strokes, The 278
Sun Ra 250

Tangerine Dream 255, 267
Tanner, Gid 216
Terry, Sonny 216, 224
Tharpe, Sister Rosetta 204
Tillman, Charles D 204
Tokio 278
Tosh, Peter 262, 263
Toussaint, Allen 264
Travis 264
Tsutsumi, Kyōhei 278
Tubb, Ernest 192
Tympany Five, The 253

Valens, Ritchie 258
Van Vliet, Don (see Captain Beefheart)
Vance, Alfred 30
Vocal Sampling 256, 274

Waits, Tom 255, 267, 269
Wallace, Wesley 245-6
Waller, Fats 252
Warren, Harry 252, 271
Washboard Sam 224
Waters, Muddy 199, 221, 226, 259

Weatherley, Jim 262
White, Booker 246
Whitter, Henry 234, 240
Wilkins, Rev. Robert 246
Wilson, Clarence 227-8
Wilson, James T 203
Winters, Lawrence 259
Wood, Roy 262
Work, Henry Clay 234
Workingmen, The 262
Wyatt, Robert 266

Yes 264

Zappa, Frank 267-8

Index of Song Titles

SONGWRITERS, ARTISTS AND their album titles will be found in the 'Popular music index' under the artist's name. Works by classical music composers will be found in the 'Index of composers and their works' under the composer's name.

9.20 Special 252

A little idea of my own 30
A new song on the opening of the Birmingham and Liverpool Railway 6
Across the track blues 248
All change for Llanfairfechan 29
Ambitious Japan 278
And the leaves began to fall 30
Automatic stop 278
Awful catastrophe at the Clayton Hill Tunnel on the Brighton Railway 6
Awful railway accident between Peterborough and Huntingdon 6
Awful railway accident, breaking of a bridge over the River Tay 6

Ballad of Casey Jones, The 210
Bang went the chance of a lifetime 30
Battle of the navvies 5
Because he was only a tramp 192, 193
Bellerin' Plain 268
Big boy blues 246
Big Rock Candy Mountain 195, n. 206
Bold English navvy 5
Boogie-woogie choo choo train 252
Brakeman's blues, The 197
Build that railroad (sing that song) 248
Buy your ticket over Rock Island lines 228
Bye bye black smoke choo choo 252

C & NW railroad blues 241
C & O excursion 240
C and NW blues 240
Casey Jones (Grateful Dead song) 255, 269-70
Casey Jones 210-13
Casey Jones, the Brave Engineer 213
Casey Jones—the Union Scab 213
Caution (do not stop on the tracks)

269
Champagne Charlie 36
Chattanooga choo choo 251-2
Cheap excursion train, The 29, 35-7
Chickasaw Special 240
Chickasaw Train blues (low down dirty thing) 200, 201-2
Choo choo 252
Choo choo blues 252
Choo choo ch'boogie 239, 252, 253
Choo choo (gotta hurry home) 248
Clang boom steam 269
Clark Gable 278
Click clack 255, 267-8
Coal train III, 255, 276-7
Cockneys trip To Brummagem, The 7
Come landlord fill the flowing bowl 5

Dashing navigator, A new comical song called the 6
Daybreak Express I, 247, 248-9
Dear old Shepherd's Bush 30
Desperados waiting for a train 262
Dixie Flyer blues 200 202, 240
Don't miss that train 262
Don't stick it out like that 30
Donkey on the railroad track 241
Double headed train 240
Down there by the train 269
Downtown train 269
Dreadful railway accident to the Irish Mail, The 6
Duquesne whistle 256, 270
Dying hobo, the 192

East coast racer 277
El Tren 274
Enfilade 278
Engineer Frank Hawk 241

Fall of Tay Bridge 6
Falling of nine arches and fifteen lives lost at Ashton, The 6
Fast freight 257
Fast movin' train 257
Fast train 257
First train home 259
Fast train through Arkansas 257
Flyin' Crow 224-5, n. 235
Freight train 225-6
Freight train blues 222, 225, 227-8, 256, 270
Freight wreck at Altoona, The 231
Frisco leaving Birmingham 246

Girl on a train 278
Give me a ticket to heaven 29
Glasgow and Ayr Railway II, 6
Glasgow is improving daily 6, 17
Going home train 259
Going North 246
Goodbye my lover goodbye 228
Gospel train 269
Grandfather's clock 234
Great Rock Island Route, The 222
Great Western Railroad or, the pleasures of travelling by steam, The 6

Halifax, Thornton and Keighley Railway 6
He's gone where they don't play billiards 30
Hobo Bill's last ride II, 192, 196, 197
Hobo blues 192, 194
Hobo's lullabye 192
Hobo's meditation 192, 197
Home in a boxcar 259
Homebound train 259

Homeward bound 255, 258, 259-60
Honky tonk train blues 244
Hull and Holderness Railway 5
Husband's boat, The 30

I been to Georgia on a fast train 257
I couldn't get in! 29
I hate that train called the M&O 200-1
I kept on waving my flag 29
I looked out of the window 30
I took the last train 264
I want you to notice my leggings 30
I'll be home on Christmas day 259
I'll keep it with mine 270
I'm a train 256, 271
I've never lost my last train, yet 31, 32
Irish harvestmen's triumph 5
Irish navigator, The 50
It ain't all honey and it ain't all jam 29
It takes a lot to laugh it takes a train to cry 270

Jack Straw 269
Jesse James 207, 216-18
Jessie at the railway bar 7
Jim Crow's description of the New Greenwich Railroad 6
John Henry 191, 207-10, 229, 277
Johnny Green's trip fro' Owdhum to see the Manchester Railway 6
Johnny the engine driver 29
Just missed the train 262

Kassie Jones 213
Kendal Fair 6, 14-15
Kiss in the railway, The 30

Last chance train 264

Last steam engine train 264
Last train 263, 264
Last train (Arlo Guthrie) 263
Last train (Captain Sensible) 265
Last train home 259, 264
Last train to Awesometown 264
Last train to Clarksville 263, 264
Last train to London 264
Let's go round and have a taster 30
Life's railway to Heaven III, 204-5
Lightnin' 248
Lines on the railway collision at Burscough Junction 6
Little black train 205
Liverpool improving daily 6, 18
Liverpool's an altered town 6, 18
Loco Madi 248
Loco-Motion, The 255, 256-7
London, Chatham and Dover 30
Lost her way 31
Lost luggage man, The 29
Lost train blues 240, 241
Love in vain 198, 199-200

McAbee's railroad piece 240
Manchester's an altered town 6, 18
Marrakesh Express 272-3
Midnight Special II, 222-4
Midnight train to Georgia 255, 262, 279
Morningtown ride II, 256, 272
Muddle Puddle Porter, The 7, 29

Navigators, The 6
Navvy boy, The 4, 5
Navvy on the line I, 5, 9-10
Navy boys, The 6
Newcastle and Carlisle Railway. A new song 4, 6, 19

New London Railway, The 6
New lost train blues 241
New Market wreck, The 230
New potato caboose 269
Night train to Memphis 222
No. 29 245-6

Oh blow the scenery on the railway 30
Oh! Mister Porter 29, 31-4
Oh! Mister what's-er-name 31
Old circus train turn-around blues,
    The 247, 248
Opening of the new railway, The 6
Opening the Oxford Railway 6
Orange Blossom Special 240, 241

Paddy one day from Greenock Town 5
Paddy on the railway 5
Paddy works on the railway 11-12
Pan American blues 240
Panama Limited 246
Peanut Special 241
Pennyworth of fun 6
Pistol Pete's Midnight Special 224
Polly put the kettle on 228
Poor Paddy works on the railway 12

Ragtime Texas 246
Railroad Bill 207, 214-16
Railroad blues 240
Railroad boomer, The 192
Railroad tramp 192
Rail, the rail, The 4, 6
Railway belle and the railway guard,
    The 29
Railway fireman, The 29
Railway guard, The 29, n. 43
Railway King, The 6
Railway Mania 6

Railway porter, Dan 29, 37-8
Railway station sandwich, The 30
Railway, The 6
Rambling on my mind III, 198
Riding in a railway train 7
Riding the blinds 240
Rock Island Line 227-30
Roots train 273
Rosie had a very rosy time 29
Runaway train, The 256, 271-2

Shillibeers Original Omnibus versus
    the Greenwich Railroad 4
Ship that never returned, The 234
Show me a train that goes to London
    30
Shunting pole inspector 7
Signalman on the line 7
Sleepy Town train 252
Slow freight 252
Slow train (Dylan) 258, 270
Slow train (Flanders and Swann) II,
    257-58
Slow train to nowhere 258
Smokestack lightnin' 265
Southern Cannonball 197
Southern Casey Jones 213
Southern No 111 241
Special streamline 246
Station to station 267, 274
Stop that train 262, 263
Streamline 'Frisco Limited 246
Sunshine Special 222

Take me back to dear old Blighty 30
Take the A Train 247, 248, 249-50,
    252
Take this hammer II, 189-91
Talking Casey blues 247

INDEX OF SONG TITLES

Terrapin station 269
There's danger on the line. The great semaphore song 30
Tickle me Timothy do 30
Tiger Rag 248
Tommy make room for your uncle 31
Tons of steel 269
Train A-Travelling 270
Traineater 277
Train fare home 259
Train home 259
Train ride home 259, 278
Train song 269
Train Special 241
Train that's taking you home, The 30
Train whistle blues 197
Trains and boats and planes 258
Trans-Europe Express 273-4
Travelling blues 246
Tuppenny Tube, The 30
Tuxedo Junction 252
Twist, The 256

Underground railway, The 30

Wabash Cannonball, The II, 192, 221-2
Waitin' for the train to come in 262
Waiting for a train (wild and reckless hobo) 192
Waiting for a train 197, 262
Waiting for the '103' 262
Walking blues 198, 199
Watching the trains come in 30
Watching the trains go out 30
Wave the flag and stop the train 262
Western Kentucky Limited 241
Western Railroad 4, 6
What did she know about railways? 31

Wheeltapper's song, The 29
When the midnight choo-choo leaves for Alabam 252
Where do flies go in the winter time? 30
Wonderful effects of the Leicester Railroad 6
Woody and Dutch on the slow train to Peking 258
Wreck of Number Four, The 231
Wreck of Number Nine, The 231
Wreck of the 1256, The 231
Wreck of the 1262, The 231
Wreck of the Old Southern 97, The 234
Wreck of the *Royal Palm*, The 231
Wreck of the *Virginian Number Three*, The 231
Wreck on the C & O Number Five, The 231
Wreck on the C & O, The 230
Wreck on the Old 97, The 221, 230, 231-2

You can't punch my ticket again 30
Young man on the railway, The 29

Select Bibliography

Anthony, Scott. *Night Mail*. London: British Film Institute, 2007.

Bagwell, Philip and Lyth, Peter. *Transport in Britain. From Canal Lock to Gridlock*. London: Hambledon and London, 2002.

Bailey, Peter, ed. *Music Hall. The Business of Pleasure*. Milton Keynes: Open University Press, 1986

Baker, Richard Anthony. *British Music Hall: An Illustrated History.*Barnsley: Pen & Sword Books Ltd, 2014.

Barry, Robert. *The Music of the Future*. London: Repeater Books, 2016.

Berlioz, Hector. *The Musical Madhouse*. Translated by Alastair Bruce. Rochester: The University of Rochester Press, 2003.

Bradley, Simon. *The Railways: Nation, Network and People*. London: Profile Books, 2015.

Brown, David. *Mikhail Glinka. A Biographical and Critical Study*. London: Oxford University Press, 1974.

Bryant, Chad. "*Into an Uncertain Future: Railroads and Vormärz Liberalism in Brno, Vienna, and Prague*". Austrian History Yearbook, 40 (2009): 183–201.

Cage, John. *Silence. Lectures and writings by John Cage*. Middletown, Conn.:Wesleyan University Press, 1961.

Cohen, Norm. *Long steel rail: the railroad in American folksong*. Urbana: University of Illinois Press, 2000.

Coleman, Terry. *The Railway Navvies*. London: Hutchinson, 1965.

Copland, Aaron and Perlis, V. *Copland, Volume I: 1900–1942*. London: Faber & Faber, 1984.

Dawes, Kwame. *Bob Marley. Lyrical Genius*. London: Bobcat Books, 2002.
Denselow, Robin. *When the music's over. The story of political pop*. London: Faber & Faber, 1989. Doggett, Peter. *The Man Who Sold the World: David Bowie and the 1970s*. London: Vintage, 2012. Dorson, Richard M. 'The Career of "John Henry"'. *Western Folklore*, July, 1965, Vol. 24, No. 3. Long Beach California: Western States Folklore Society: 155-163.

Earnshaw, Alan. *The Holmfirth (Summer Wine) Branch Line*. Appleby-in-Westmorland, Cumbria: Trans-Pennine Publishing Ltd., 2005.
Ellington, Edward Kennedy. *Music is my mistress*. New York: Doubleday. 1974.

Figes, Orlando. *The Europeans: Three Lives and the Making of a Cosmopolitan Culture*. London: Penguin, 2020.
Foreman, Lewis, ed. *The Percy Grainger Companion*. London: Thames Publishing.
Fox, Christopher. 'Steve Reich's "Different Trains"' in *Tempo*, March, 1990, New Series, No. 172 (March, 1990): 2-8.

Garon, Paul. *What's the Use of Walking If There's a Freight Train Going Your Way? Black Hoboes & Their Songs*. Chicago: Charles H Kerr Publishing Company, 2006.
Gibbons, Jack. http://www.jackgibbons.com/alkanmyths.
Gioia, Ted. *Work Songs*. Durham and London: Duke University Press, 2006.
Glinka, Mikhail Ivanovich. *Memoirs*. Translated by Richard B Mudge. Norman: University of Oklahoma Press, 1963.
Gourvish, T R. *British Railways 1948-73. A Business History*. Cambridge: Cambridge University Press, 1986.
GPO Film Unit (1933-1940). http://www.screenonline.org.uk/film/.
Griffiths, Paul. *A Concise History of Modern Music. From Debussy to Boulez*. London: Thames and Hudson, 1978.
Grove Music Online. https://www-oxfordmusiconline.

Hammond, Nick and Keygnaert, Frederick. "Le Chemin de Fer". *The Alkan Society*, Bulletin no. 96 (April 2018): 14-16.
Herbert, Trevor and Myers, Arnold. "Music for the multitude: accounts of brass bands entering Enderby Jackson's Crystal Palace contests in the 1860s." *Early Music* 38, no. 4 (2010): 572-84.
Herbert, Trevor. *Bands: The Brass Band Movement in the 19th and 20th Centuries*. Milton Keynes: Open University Press, 1991.
Hilburn, Robert. *Paul Simon. The Life*. London, Simon and Schuster UK, 2018.

Hobsbawm, Eric. *Industry and Empire: From 1750 to the Present Day*. London: Penguin, 1969.
Honri, Peter. *Working the Halls*. London: Futura Publications, 1974.

Ives, Charles E. *Charles E Ives. Memos*. New York: W. W. Norton & Company, 1972.

Jackson, Pat. *A Retake Please! Night Mail to Western Approaches*. Liverpool: Liverpool University Press, 1999.
Jewell, Derek. *A Portrait of Duke Ellington*. London: Sphere Books, 1977.
Johnson, Julian. *Out of Time: Music and the Making of Modernity*. Oxford: Oxford University Press, 2015.

Keating P J. *The Working Class in Victorian Fiction*. London: Routledge and Kegan, 1971.
Kemp, Peter. *The Strauss Family. Portrait of a Musical Dynasty*. Tunbridge Wells: The Baton Press, 1985.
Kingsford, P W. *Victorian Railwaymen. The emergence and growth of railway labour 1830-1870*. London: Frank Cass, 1970.

Leigh, Spencer. *Simon & Garfunkel. Together Alone*. Carmarthen: McNidder and Grace. 2016.
Loft, Charles. *Government, the Railways and the Modernization of Britain: Beeching's Last Trains*. Abingdon: Routledge, 2006.
Lomax, Alan. *The Penguin Book of American Folk Songs*. London: Penguin, 1964.
Lomax, Alan. *The Folk Songs of North America*. New York: Doubleday, 1975.

Mellers, Wilfrid. *Percy Grainger*. Oxford: Oxford University Press. 1992.
Mellers, Wilfrid. *Francis Poulenc*. Oxford: Oxford University Press. 1993.
Milburn, George. *The Hobo's Hornboook*. New York: Ives Washburn, 1930.
Mitchell, Donald. *Britten and Auden in the Thirties: The Year 1936*. Woodbridge, Suffolk: Boydell Press, 1981.
Murray, Albert. *Stomping the Blues*. Minneapolis: University of Minnesota Press, 1976.

National Railway Museum https://www.railwaymuseum.org.uk/research-and-archive/further-resources.
Nelson, Scott Reynolds. *Steel Drivin' Man: John Henry, the Untold Story of an American Legend*. Oxford: Oxford University Press, 2006.

Nichols, Roger. *The Harlequin Years: Music in Paris 1917-1929*. London: Thames and Hudson, 2002.

Oliver, Paul. *Blues Fell This Morning: Meaning in the Blues*. Cambridge: Cambridge University Press, 1960.

Oliver, Paul. *Screening the Blues. Aspects of the blues tradition*. New York: Da Capo Press, 1968.

Pegg, Nicholas. *The Complete David Bowie*. London: Titan Books, 2016.

Pirie, Gordon. "Brutish bombelas. Trains for migrant gold miners in South Africa c. 1900-25". *Journal of Transport History* 18 no. 1 (1997): 31-44.

Polenberg, Richard. *Hear My Sad Story: The True Tales That Inspired "Stagolee," "John Henry," and Other Traditional American Folk Songs*. New York: Cornell University Press, 2015.

Raulerson, Graham. 'Hoboes, Rubbish, and "The Big Rock Candy Mountain"'. *American Music*, Vol. 31, No. 4, 2013: 420-449.

Revill, George. *Railway*. London: Reaktion Books Ltd., 2013.

Richards, Jeffrey and MacKenzie, John M. *The Railway Station: A Social History*. Oxford: Oxford University Press, 1986.

Rolt, L T C. *Red for Danger. The Classic History of British Railway Disasters*. Stroud: The History Press, 2009.

Russell, Dave. *Popular music in England. 1840-1914: A social history*. Manchester: Manchester University Press, 1987.

Sandburg, Carl. *The American Songbag*. New York: Harcourt Brace Jovanovich, 1927.

Schafer, R Murray. *The soundscape. Our sonic environment and the tuning of the world*. Rochester, Vermont: Destiny Books, 1977.

Scowcroft, Philip. "Railways in Music". http://www.musicweb-international.com/railways_in_music.

Simmons, Jack, ed. *Railway Traveller's Handy Book of 1862*. Bath: Adams & Dart, 1971.

Simmons, Jack. *The Victorian Railway*. London: Thames & Hudson, 1991.

Simmons, Jack and Gordon Biddle ed., *The Oxford Companion to British Railway History*. Oxford: Oxford University Press, 1997.

Southern, Eileen. *The Music of Black Americans. A History*. New York: Norton, 1971.

Stewart, John Lincoln. *Ernst Krenek. The man and his music.* Los Angeles: University of California Press, 1992.
Street, John. *Music and politics.* Cambridge: Polity Press, 2012.

Wade, Stephen. *The Beautiful Music All Around Us. Field Recordings and the American Experience.* Champaign, Illinois: University of Illinois Press, 2012.
Weissman, Dick. *Blues. The Basics.* London: Routledge, 2005.
Williams, Carolyn. *Gilbert and Sullivan. Gender, Genre and Parody.* New York: Columbia University Press, 2011.
Winterson, J, Nickol, P & Bricheno, T. *Pop Music: The Text Book.* London: Edition Peters, 2013.
Wojtczak, Helena. *Railwaywomen.* Hastings, East Sussex: Hastings Press, 2005.
Wolmar, Christian. *Blood, Iron and Gold: How the Railways Transformed the World.* Great Britain: Atlantic Books, 2010.
Wolmar, Christian. *The Golden Age of European Railways.* Barnsley: Pen & Sword Transport, 2013.
Wolmar, Christian. *A Short History of Trains.* London: Dorling Kindersley Ltd., 2019.